Constructing 3-Dimensional Models

Sham Tickoo

Associate Professor
Department of Manufacturing Engineering Technologies
and Supervision
Purdue University Calumet
Hammond, Indiana

DEDICATION

To teachers and parents, who make it possible
to disseminate knowledge
to enlighten the young and curious minds
of our future generations

To students, who are dedicated to learning
and making the world a better place to live

Published by:

CAD-CIM Technologies
525 St. Andrews Drive
Schererville, IN 46375

Tel: (219) 322-1001   Fax: 219 322 1001

NOTICE TO THE READER

Note:

The book contains 10 shapes and 8 projects. These shapes and projects are also available in the following packages:

A. Shapes (Set of 10)
B. Individual Projects
   (Project 1 through Project 8)

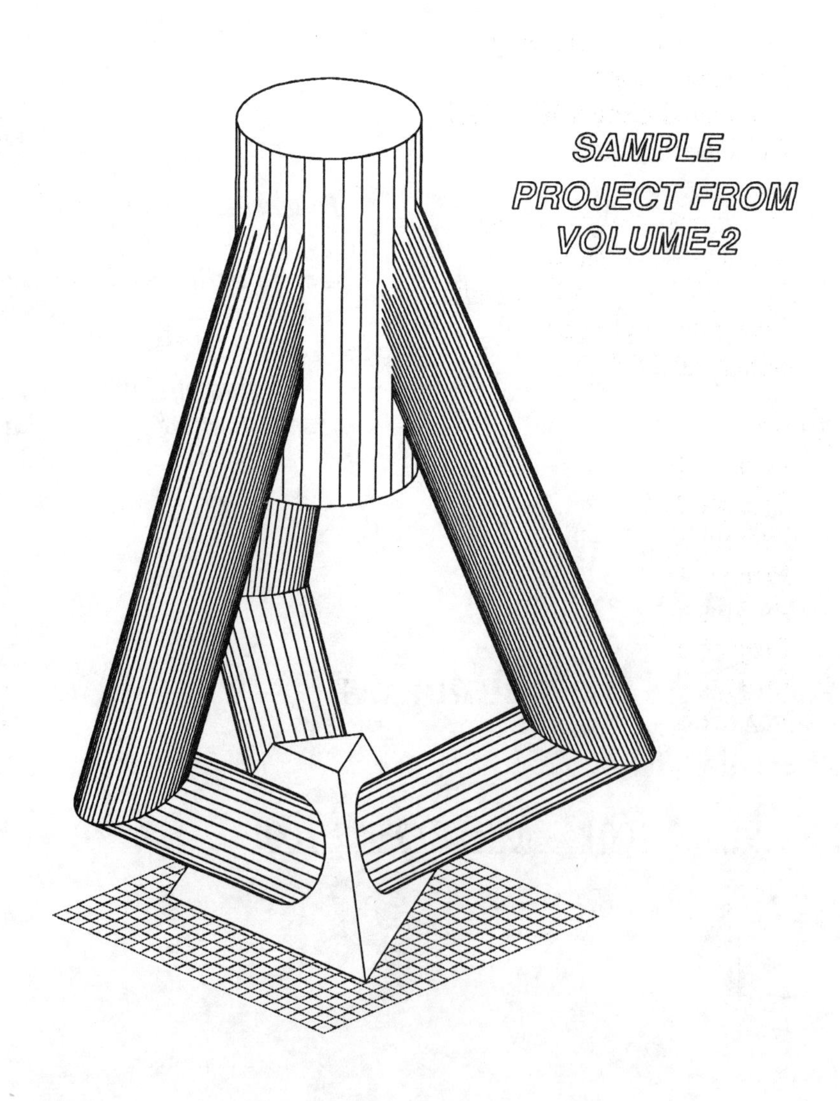

Table of Contents

Constructing 3-Dimensional Models

This book can be used in schools, colleges, and industry to learn geometrical construction techniques. In the computer aided design/drafting environment (CADD), the shapes and the projects contained in this book are excellent examples for learning 3-dimensional drawings, solid modeling, and animation. It can also be used in drafting to visualize the shapes for drawing orthographic, auxiliary, sectional, and isometric views. In industry it can be used to train sheet metal fabricators to fabricate different shapes from the flat layouts.

"They found them to be a challenge. These fifth graders worked well together despite their varying math abilities. This also works with cooperative learning. The children's comments ranged from: "This is cool", "I don't get it," "Show me how you got that to fit," "This is neat." I liked the fact that the only directions were cut, fold, match, and join. This forced my students to think as well as ask others in the group for their input"
K. Johnson, Homan Elementary School,
Schererville, IN

"The ideas in the book are great resources for class, lab. work, and for science fair projects."
Cesar Queyquep, Bishop Noll Institute,
Hammond, IN

"My children had a wonderful time putting the pieces together. They were excited when they saw the final shape."
Indira Q., Torrington, CT

"It is a new learning tool for all ages. It is amazing how complicated shapes can be formed by folding flat sheets."
Neet G., San Diego, CA

"It is fun and a good learning experience. The shapes and projects can be used to decorate Christmas trees."
Vinod S., Hammond, IN

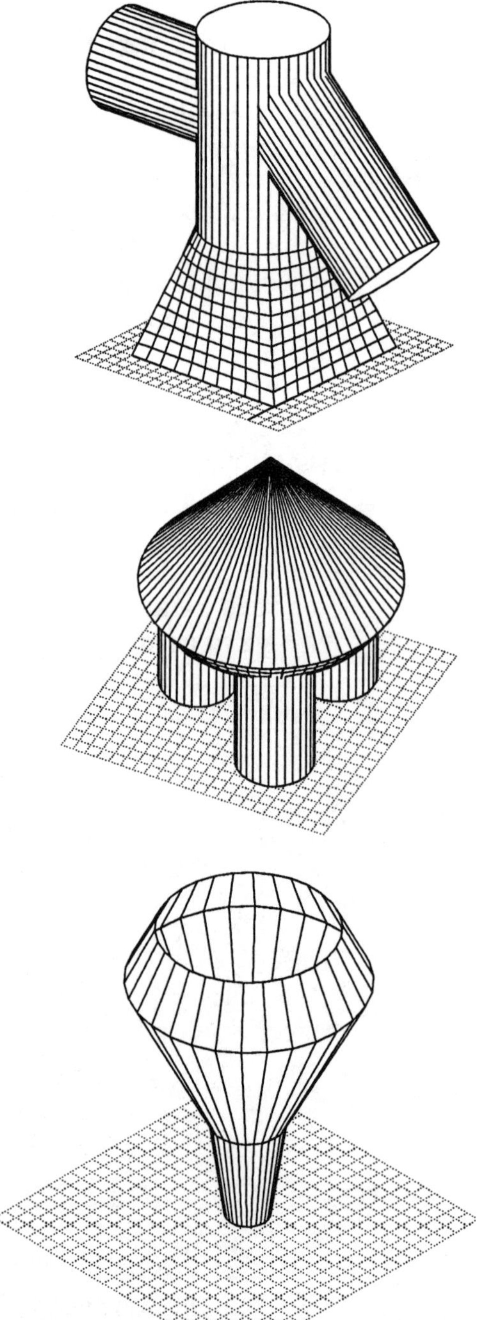

For information, please call or write to:

CAD-CIM Technologies

525 St. Andrews Drive
Schererville, IN 46375

Tel: (219) 322-1001 Fax: 219 322 1001

Preface

The purpose of this book is to give the user an idea about how a complicated shape can be generated by using simple two dimensional elements. Most of the 3-dimensional objects consist of flat and circular shapes. These basic shapes can be combined in a desired configuration to form complicated objects used in planes, ships, automobiles, appliances, agricultural and industrial equipment.

The shapes and projects that are contained in this book have been generated by using computer programs. The author of this book has developed a computer software package "SMLayout" that can be used to calculate various parameters for different shapes and draw the flat layout on the computer screen. These drawings can then be plotted or sent to a computer controlled machine tool to cut the shapes. This software is being used by several companies in North America and some countries overseas in the design and fabrication of products.

This book will help the reader to understand the process involved in making products that we see and use in everyday life. It also guarantees several days of fun filled activity. Once children get involved in building models, it will give them a sense of pride and add to their self confidence. This book can also be used by children to improve their eye-hand coordination. They will learn to follow directions to successfully accomplish a project. The projects contained in the book will provide valuable experience and education to the user.

The book can also be used in schools, colleges, and industries to practice geometrical construction techniques. In the computer aided design/drafting environment (CADD), the shapes and the projects contained in this book are excellent examples for learning 3-dimensional drawings, solid modeling, and animation. It can also be used in drafting to visualize the shapes for drawing orthographic, auxiliary, sectional, and isometric views. In industry it can be used to train sheet metal fabricators to fabricate different shapes from flat layouts.

For more information about the software "SMLayout" you should contact CAD-CIM Technologies. For information about using these shapes for 3 dimensional drawing, animation, or for solid modeling, please contact the author of this book.

We would like to thank Professor David Rose, Professor Daniel Yovich, Professor Gregory Neff, David McLees and the evaluators for their valuable suggestions.

Instructions for constructing shapes

The first volume of this work-book contains 10 shapes and 8 projects. The book also contains 3-dimensional drawings of the shapes and projects. The 3-dimensional drawings are arranged in groups, based on the similarity of shapes. These 3-dimensional drawings are followed by the flat layout of the shapes. For projects, the 3-dimensional drawings are followed by a set of flat layouts. The number of layouts depends on the complexity and the number of parts in the model.

Solid, Hidden and Dash-dot Lines

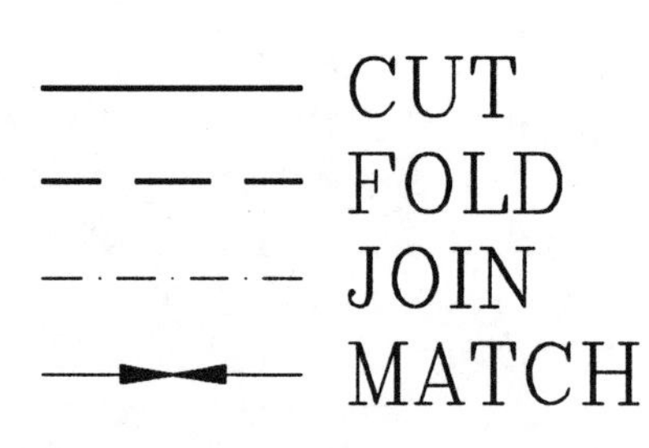

The flat layouts contain three types of lines, solid lines, dashed lines, and dash-dot lines. In addition to these lines, the layouts also use arrows with leader lines. The solid line defines the cutting line. The hidden lines indicate where the layout must be bent. If there is a partial hidden line that does not extend along the entire length, make sure that you bend the layout only along the hidden line. The dash-dot lines indicate the edges or lines that must be joined with the corresponding line to form the given shape. The arrows indicate the lines that must match. Also, the tips of the arrows should coincide to ensure correct placement of edges.

Cutting and Forming Shapes

The following steps are involved in cutting and forming the given shapes or assembling the projects.

1. Tear the sheet from the book along the perforation.

2. **Cut the layout along the solid line** using a pair of scissors. You must be careful when cutting the layout. If the layout is not cut along the line, then the parts may not fit or the corresponding lines may not match. This will result in a poor shape and the parts may not align correctly.

4. After cutting the layout you may use any color combination to color different sections of the layout.

5. Cut small pieces of tape that are approximately 3/4" x 1/4".

6. The next step is to join the edges and use the tape to secure them together. After you have joined all parts and you are satisfied with the shape, you may use glue to hold the parts together.

3D SHAPES-1

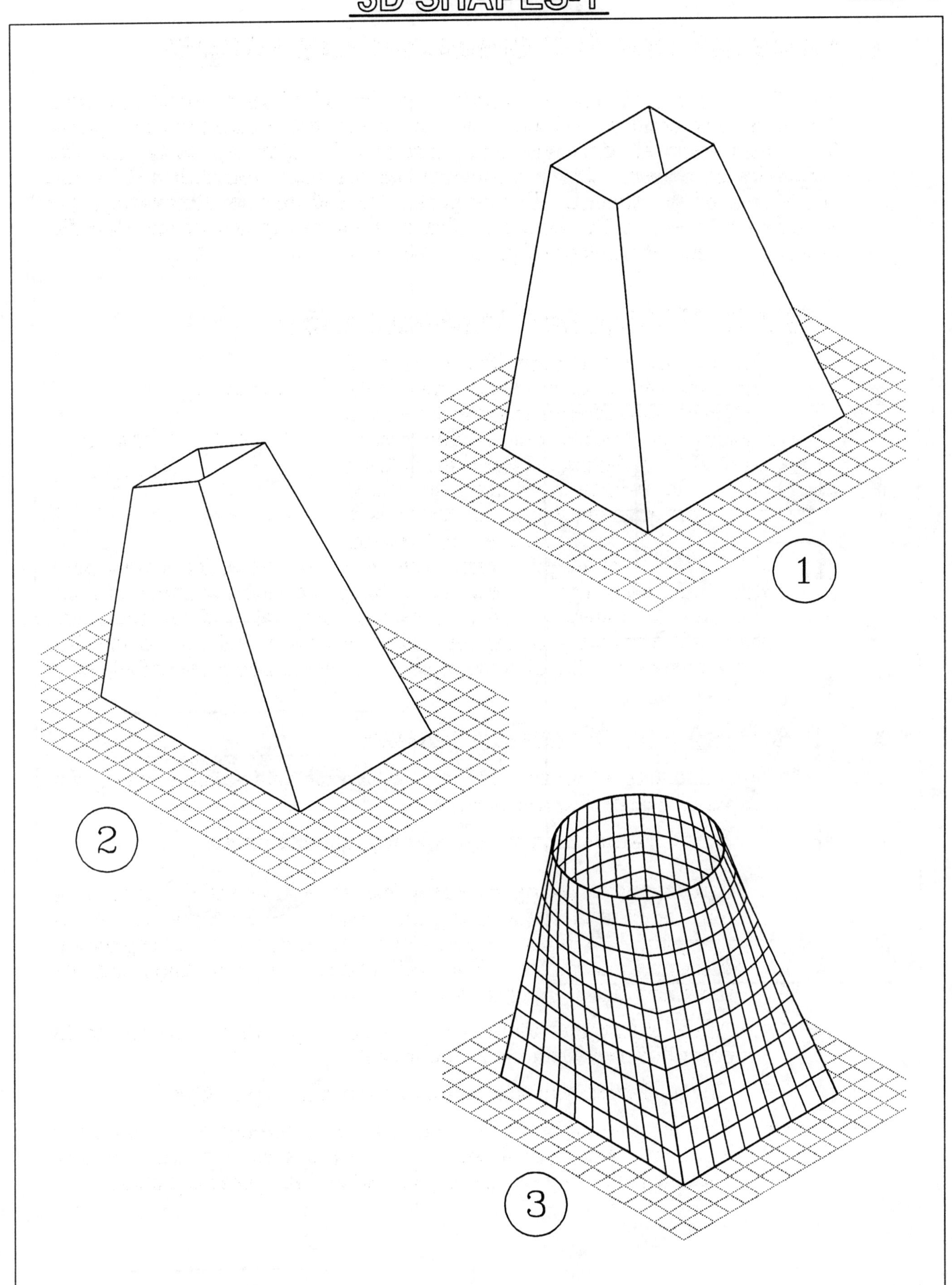

3D SHAPES-1

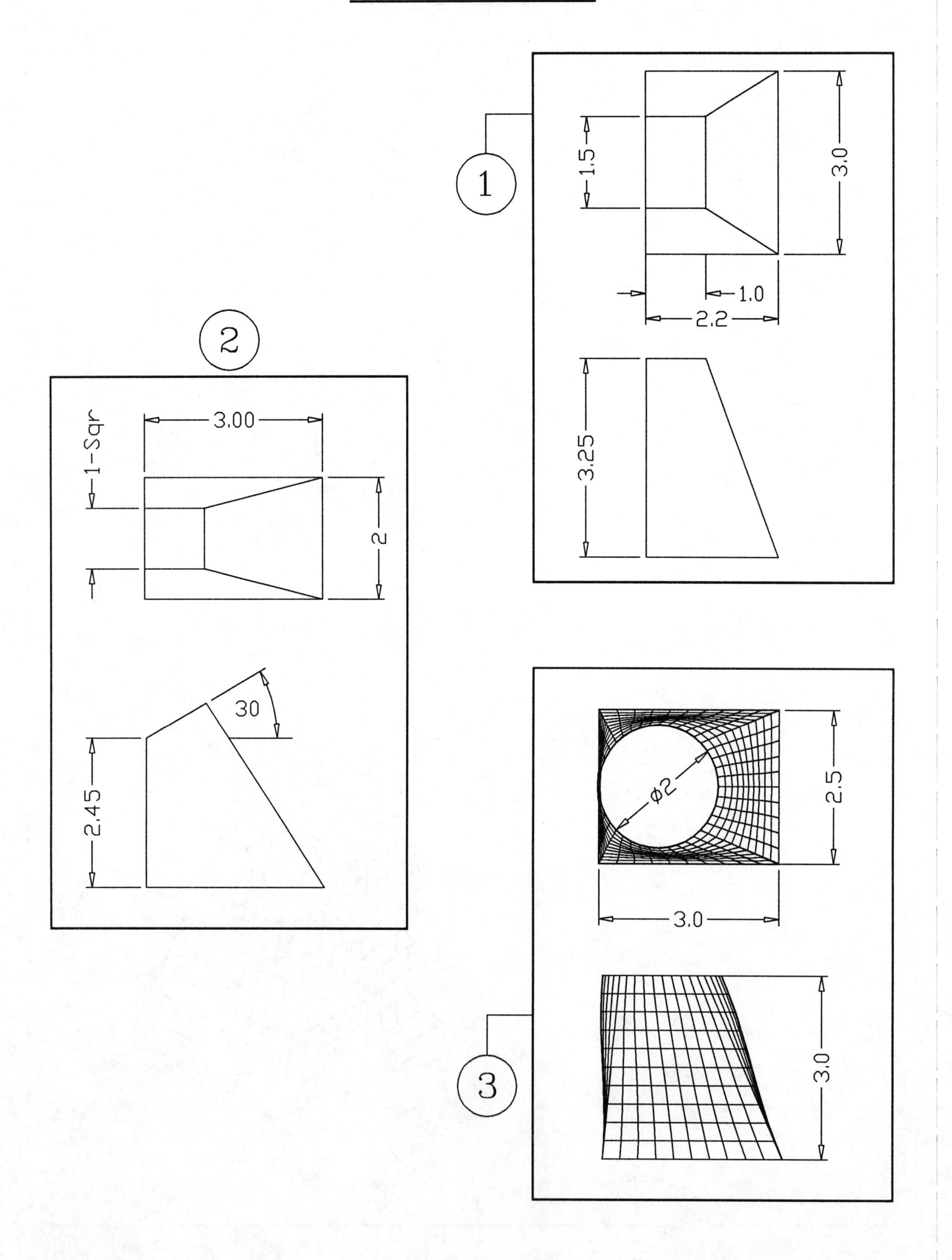

RECTANGLE TO RECTANGLE TRANSITION WITH OFFSET

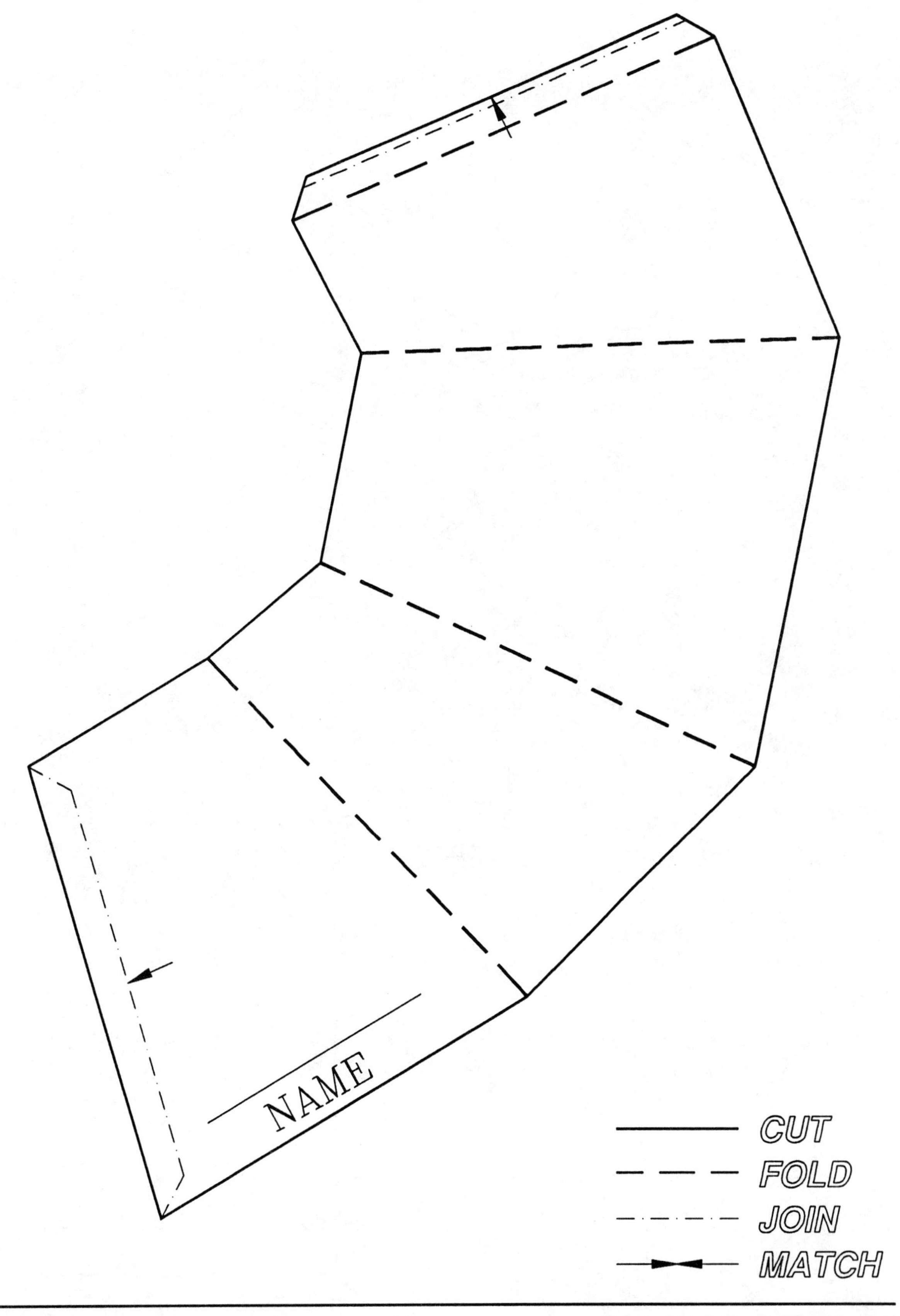

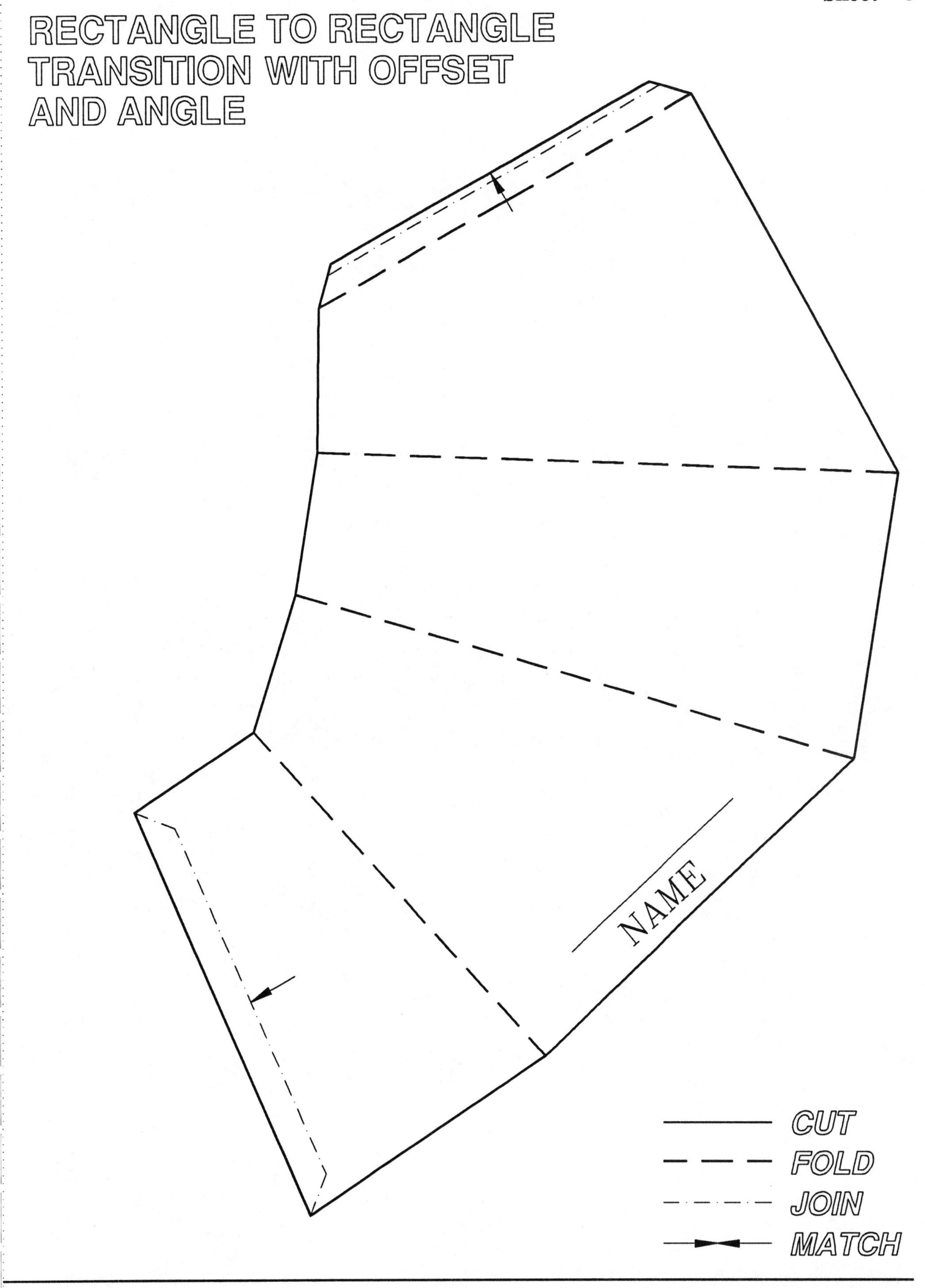
Sheet - 4
RECTANGLE TO RECTANGLE
TRANSITION WITH OFFSET
AND ANGLE
NAME
CUT
FOLD
JOIN
MATCH

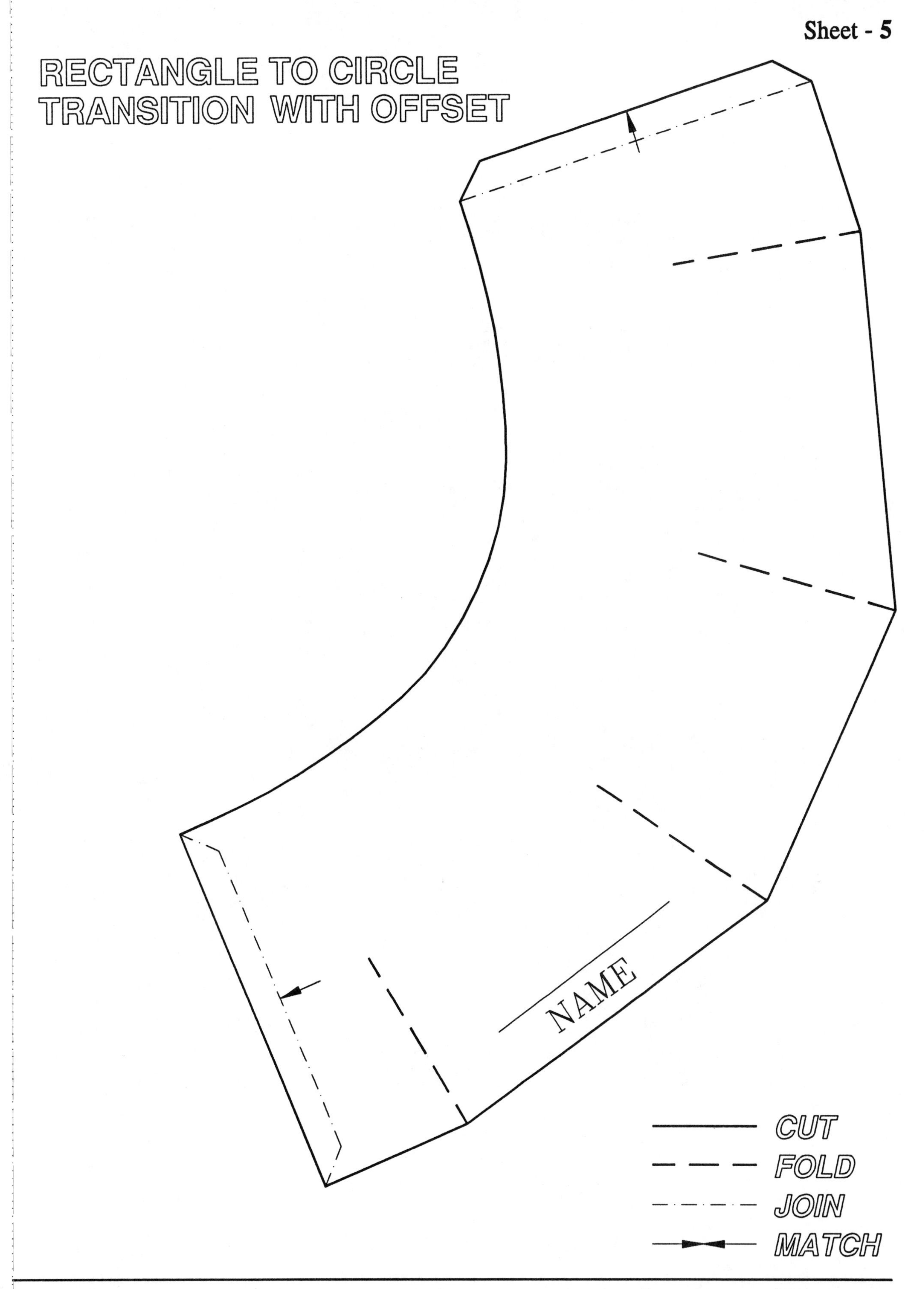
Sheet - 5
RECTANGLE TO CIRCLE
TRANSITION WITH OFFSET
NAME
CUT
FOLD
JOIN
MATCH

3D SHAPES-2

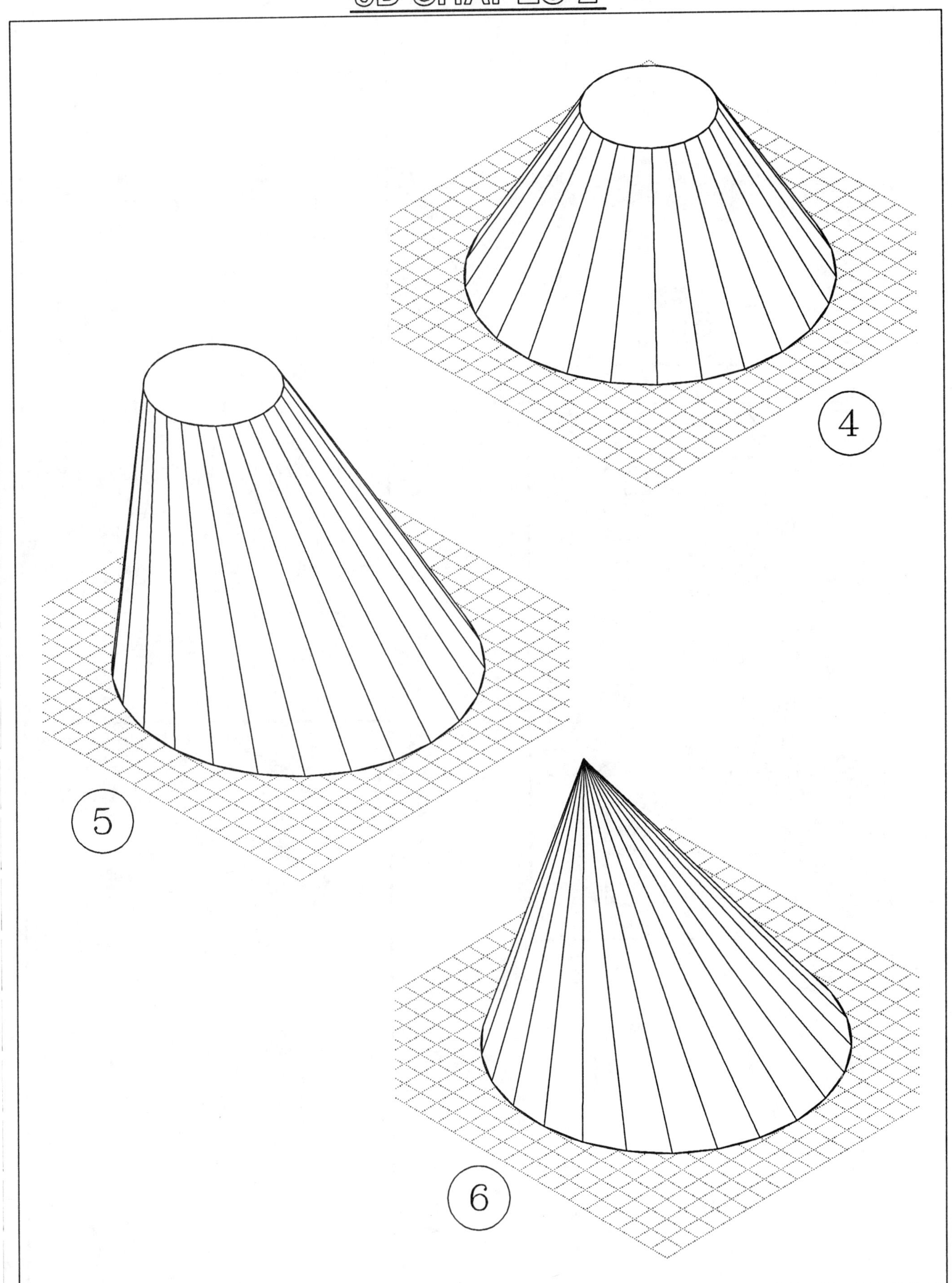

3D SHAPES-2

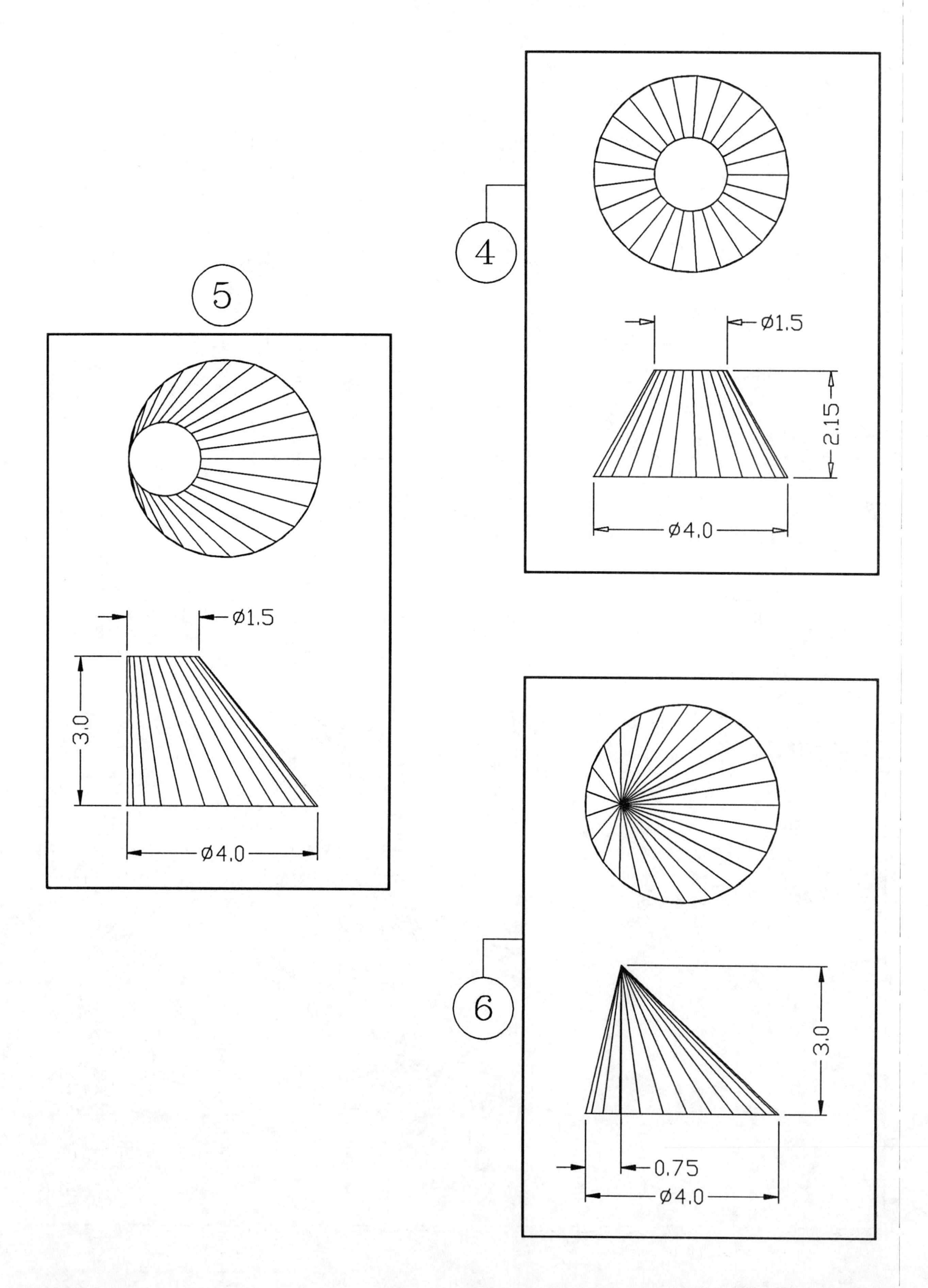

TRUNCATED CONE

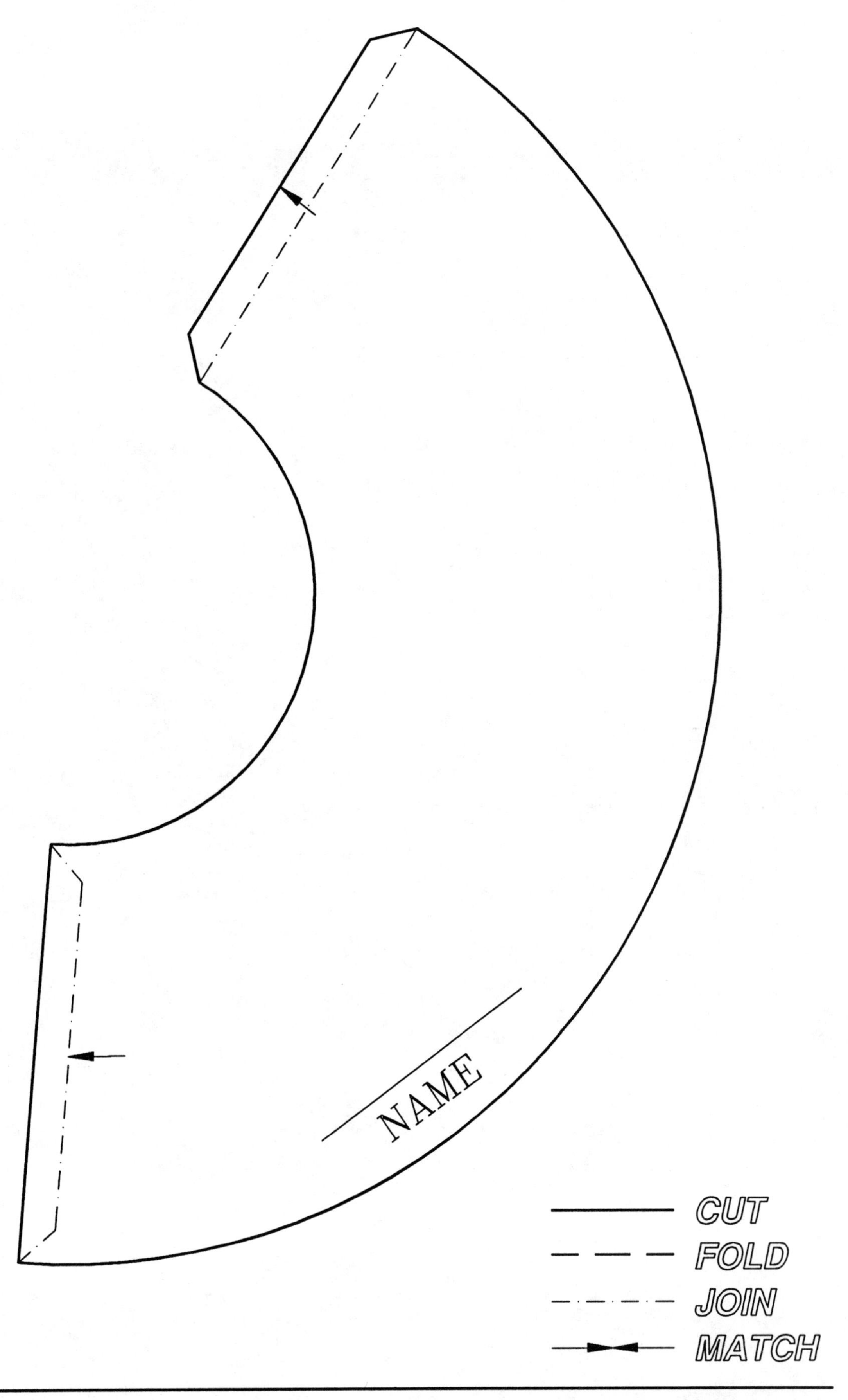

TRUNCATED CONE WITH OFFSET

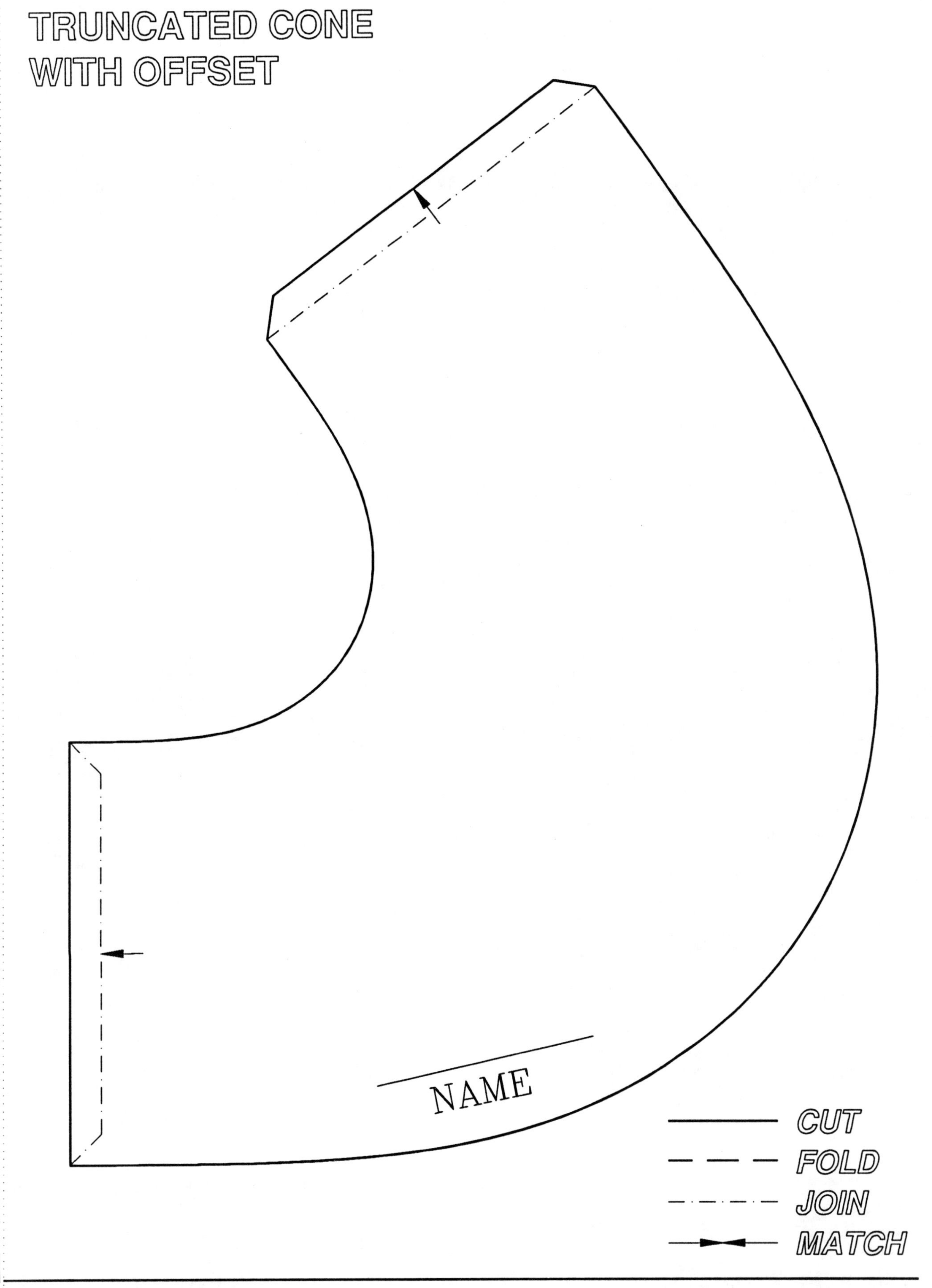

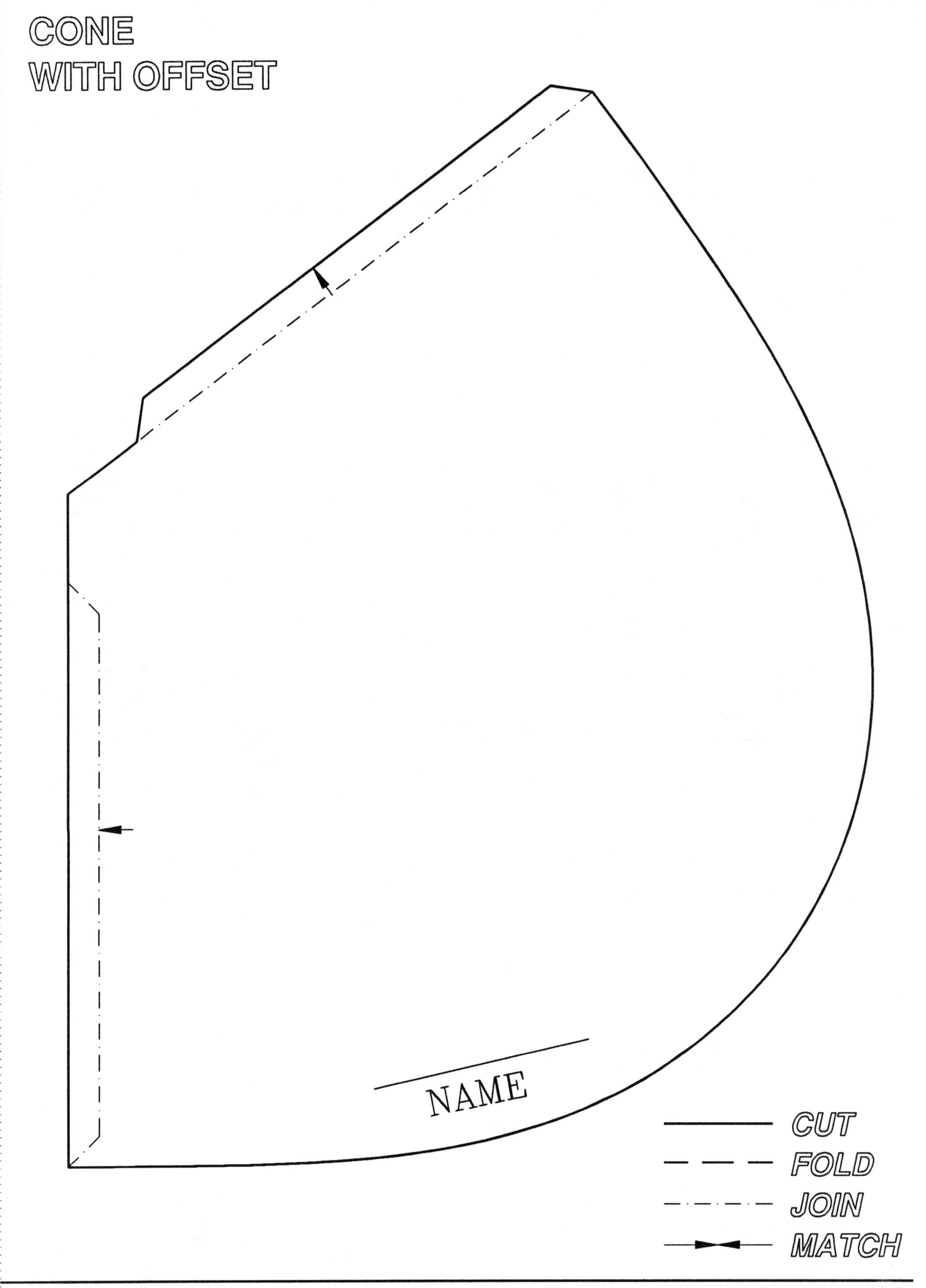
CONE
WITH OFFSET
NAME
CUT
FOLD
JOIN
MATCH

3D SHAPES-3

Sheet - 10

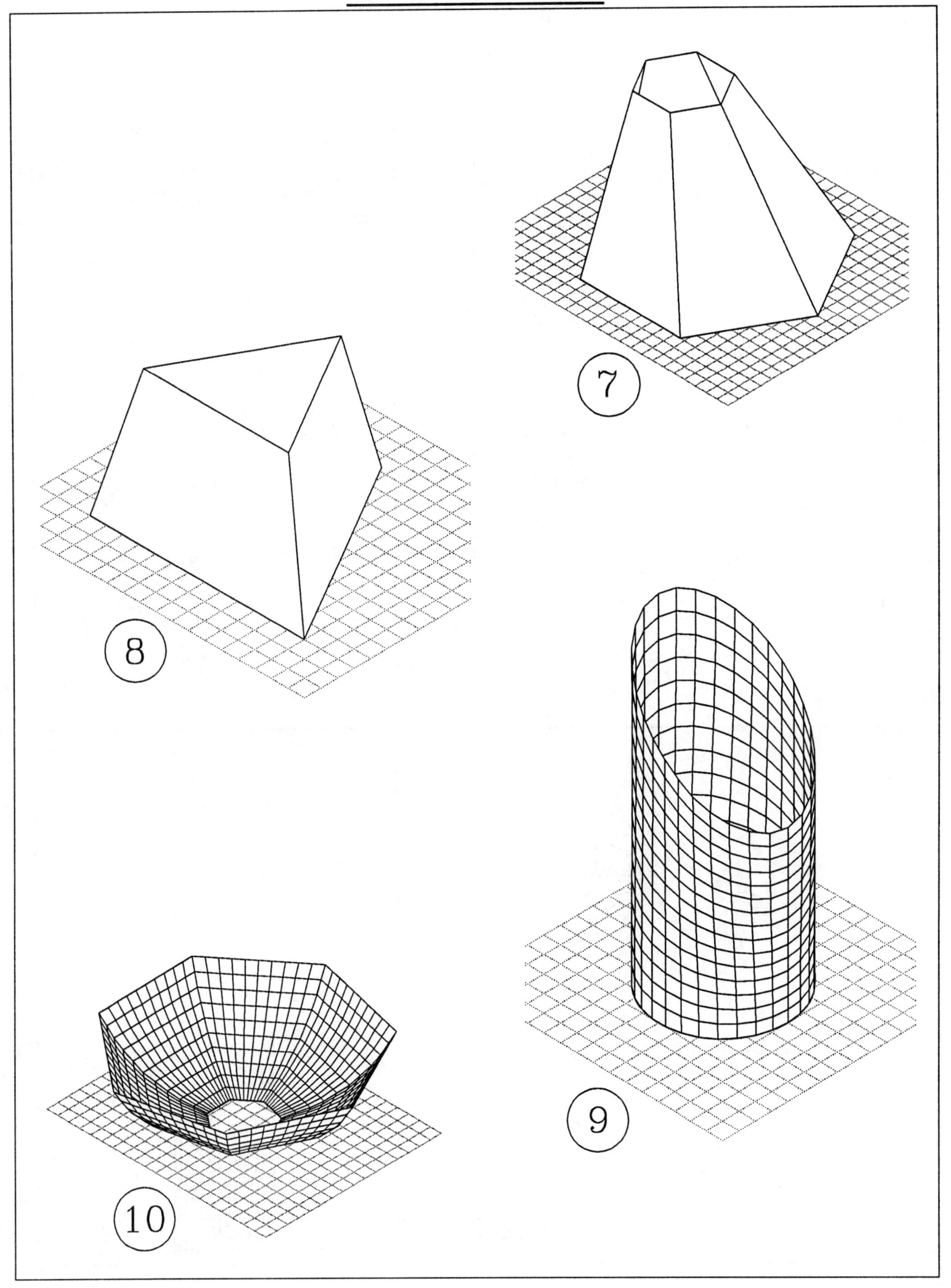

3D SHAPES-3

8

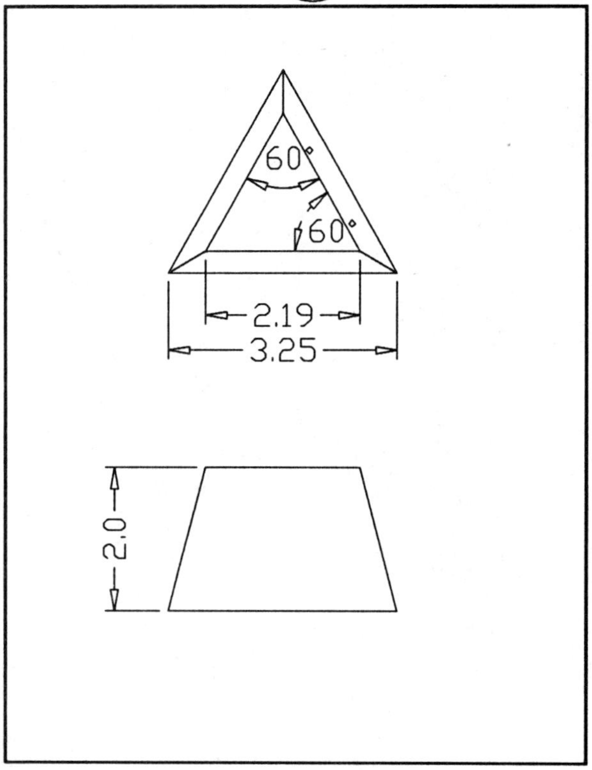

60°
60°
2.19
3.25
2.0

7

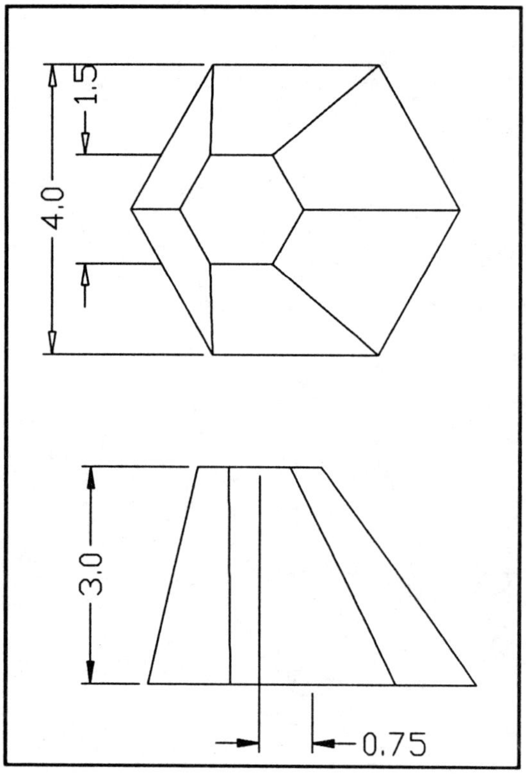

1.5
4.0
3.0
0.75

10

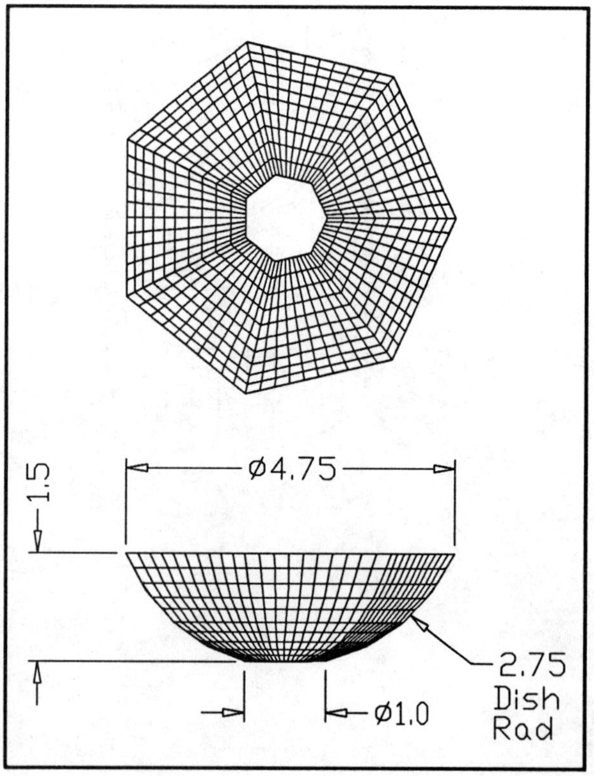

1.5
Ø4.75
2.75 Dish Rad
Ø1.0

9

Ø2.0
45
4.5

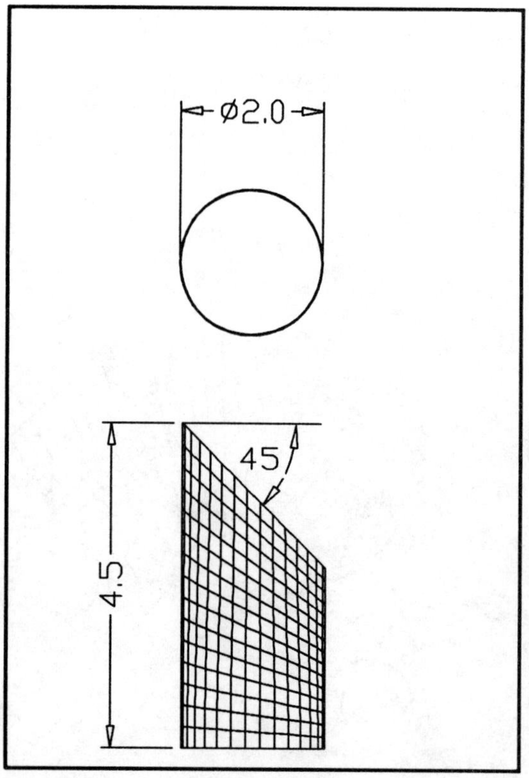

TRUNCATED PYRAMID WITH OFFSET

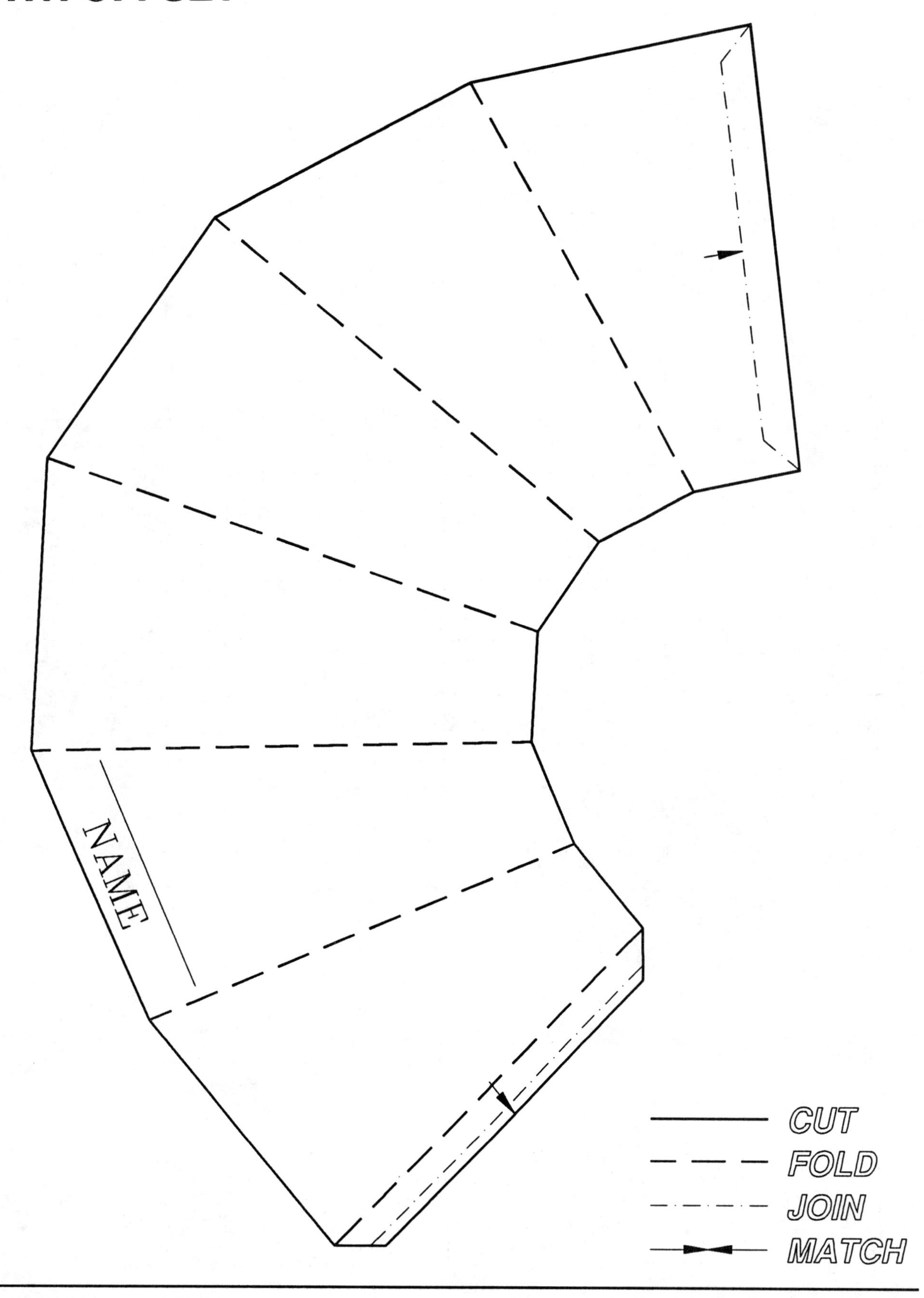

TRIANGLE TO TRIANGLE TRANSITION

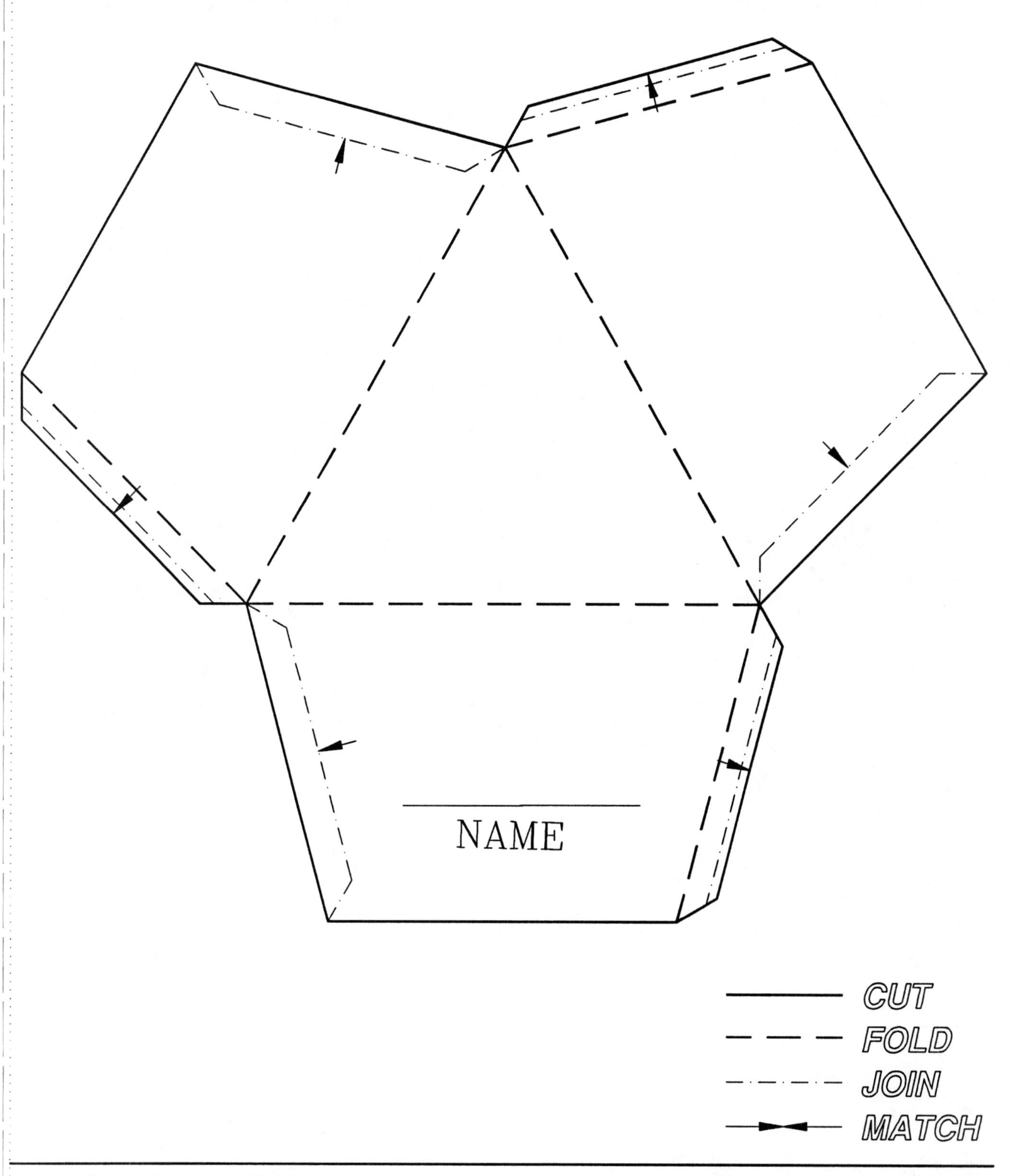

TRUNCATED CYLINDER

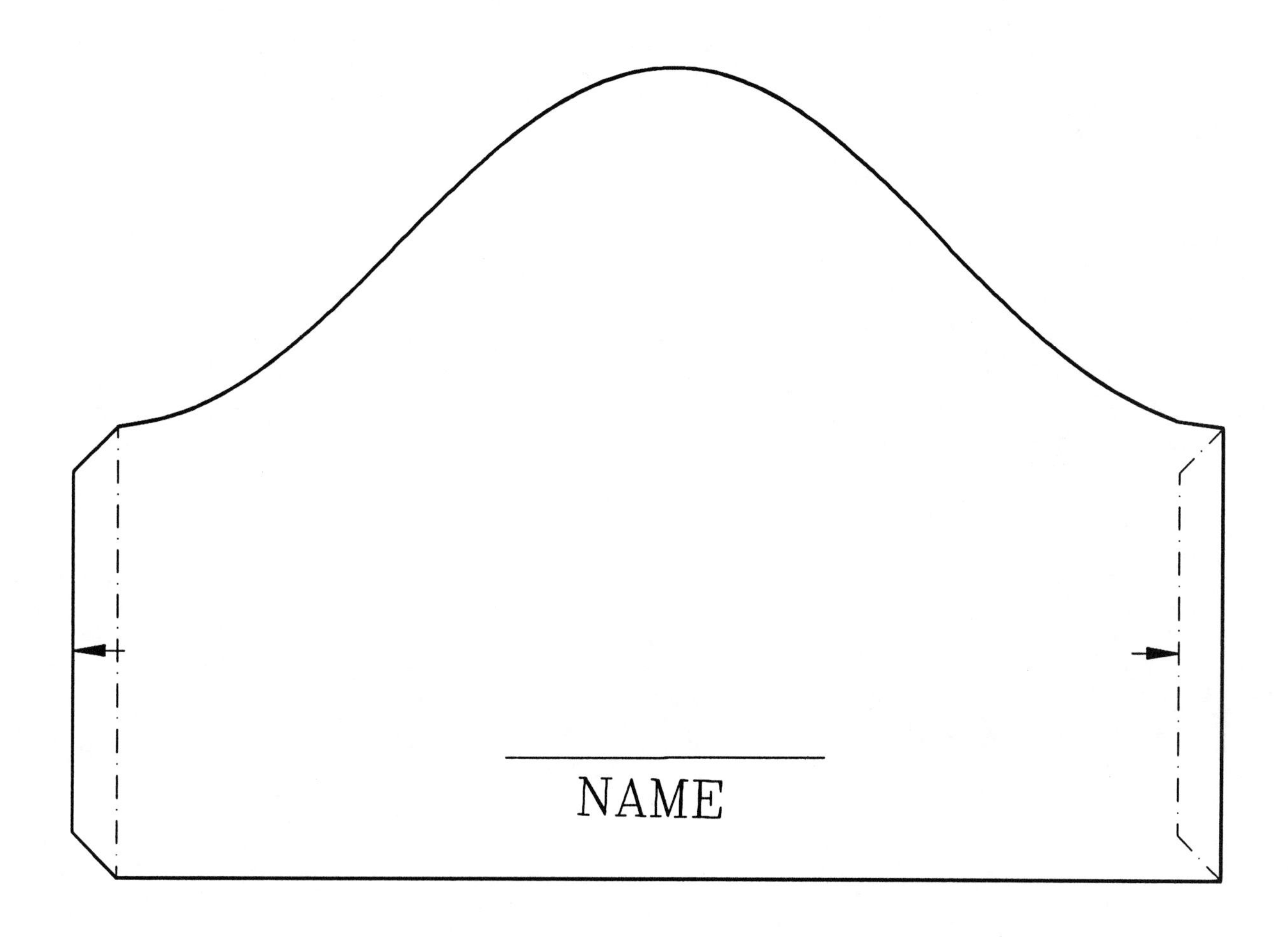

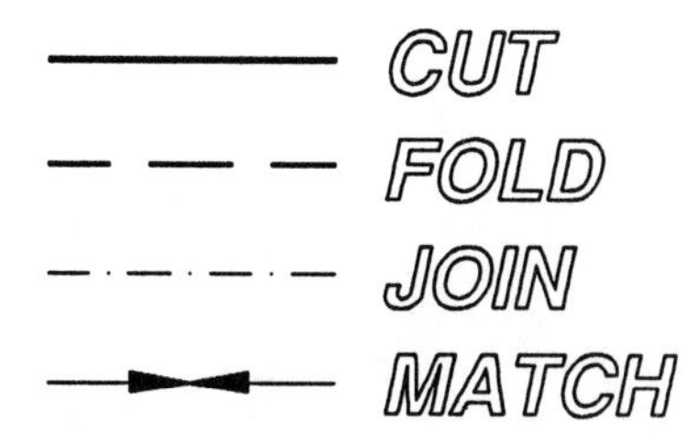

SPHERICAL DISH

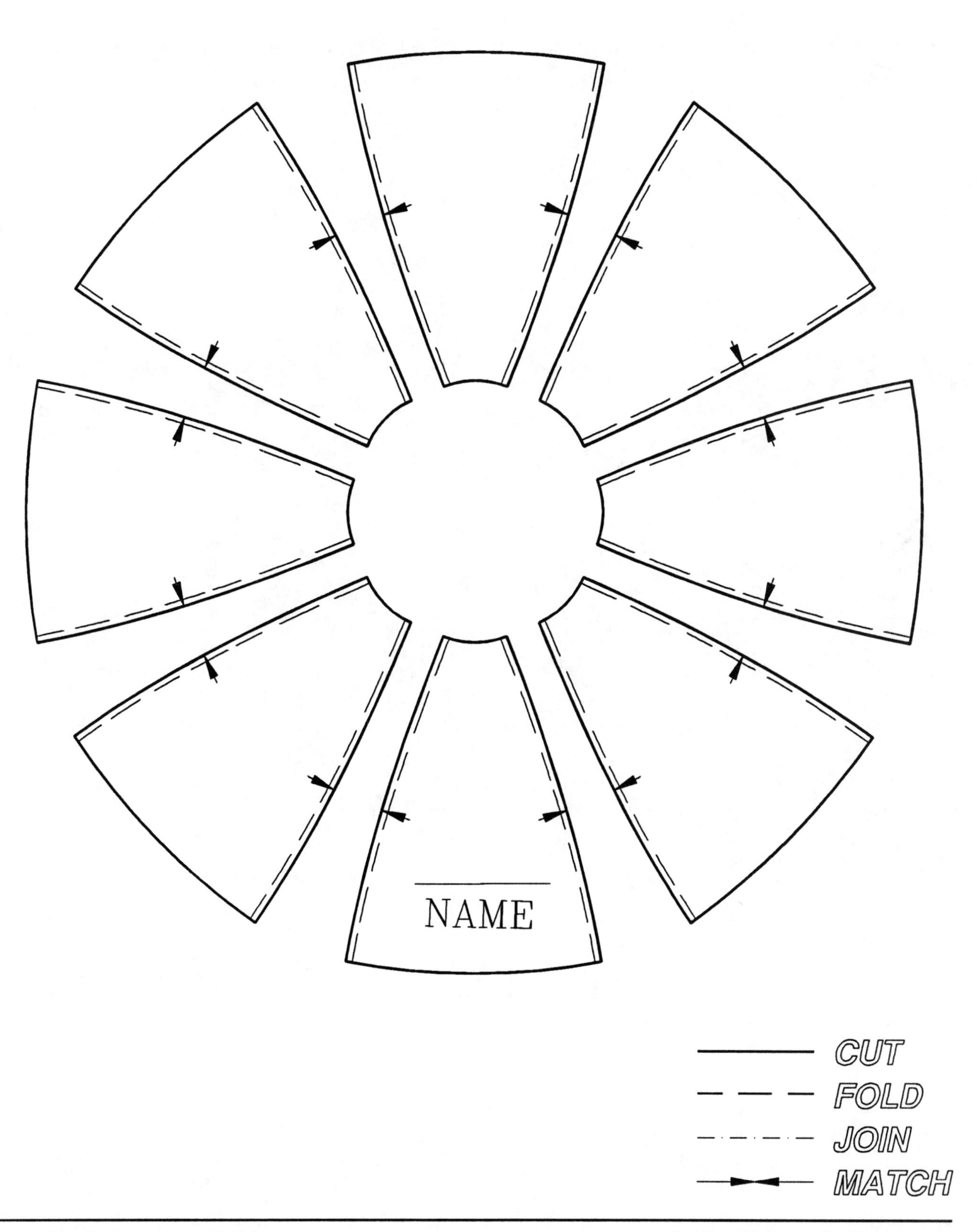

PROJECT-1

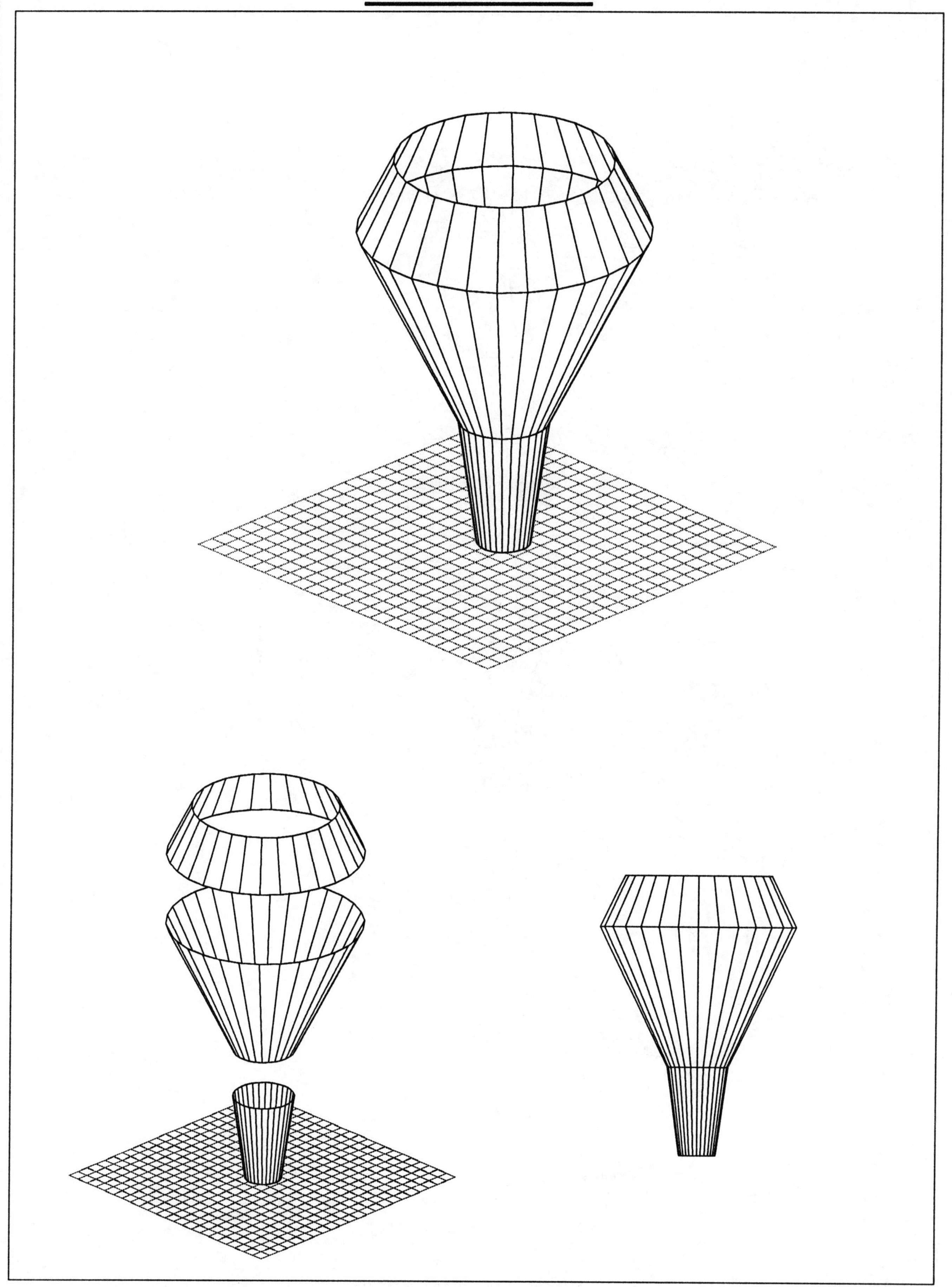

PROJECT-1

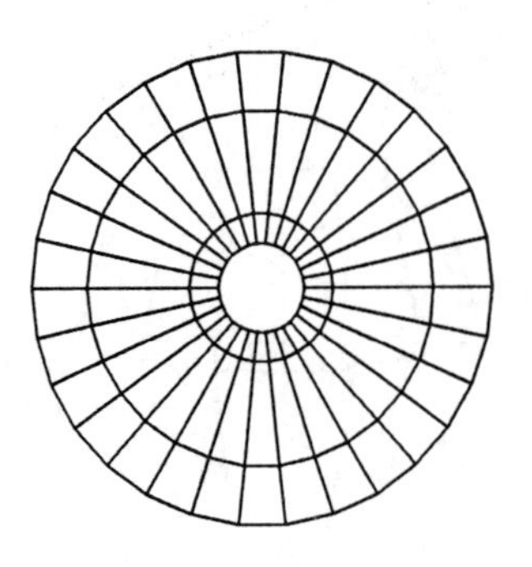

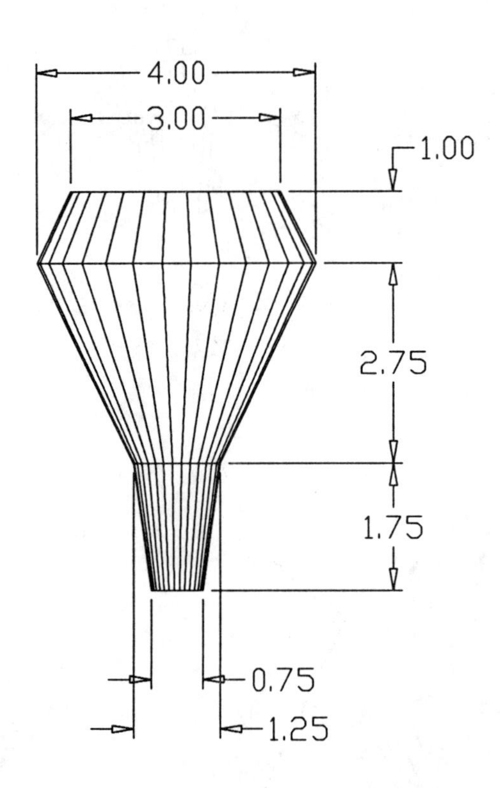

PROJECT-1

Sheet - **16**

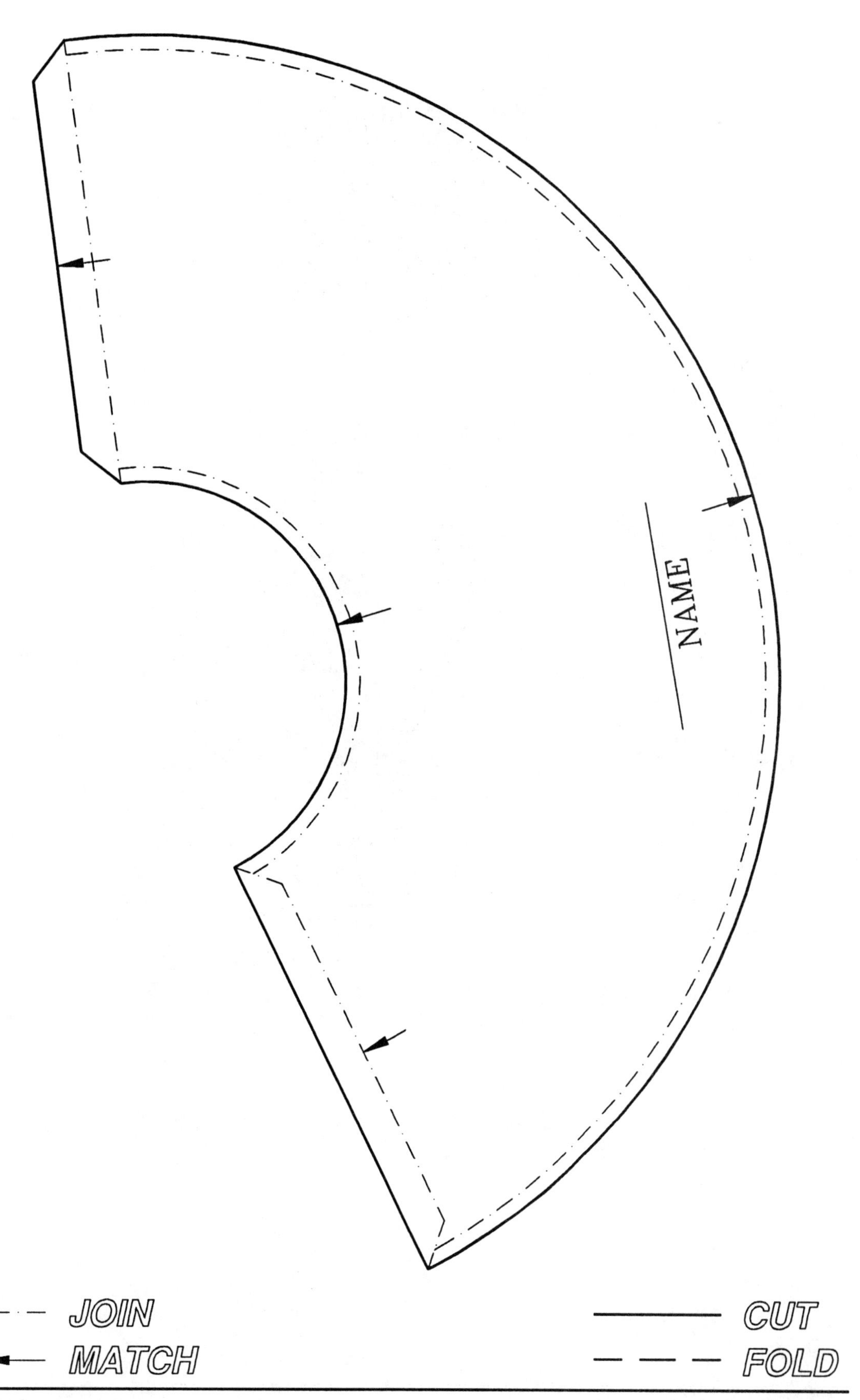

JOIN
MATCH
CUT
FOLD

PROJECT-1

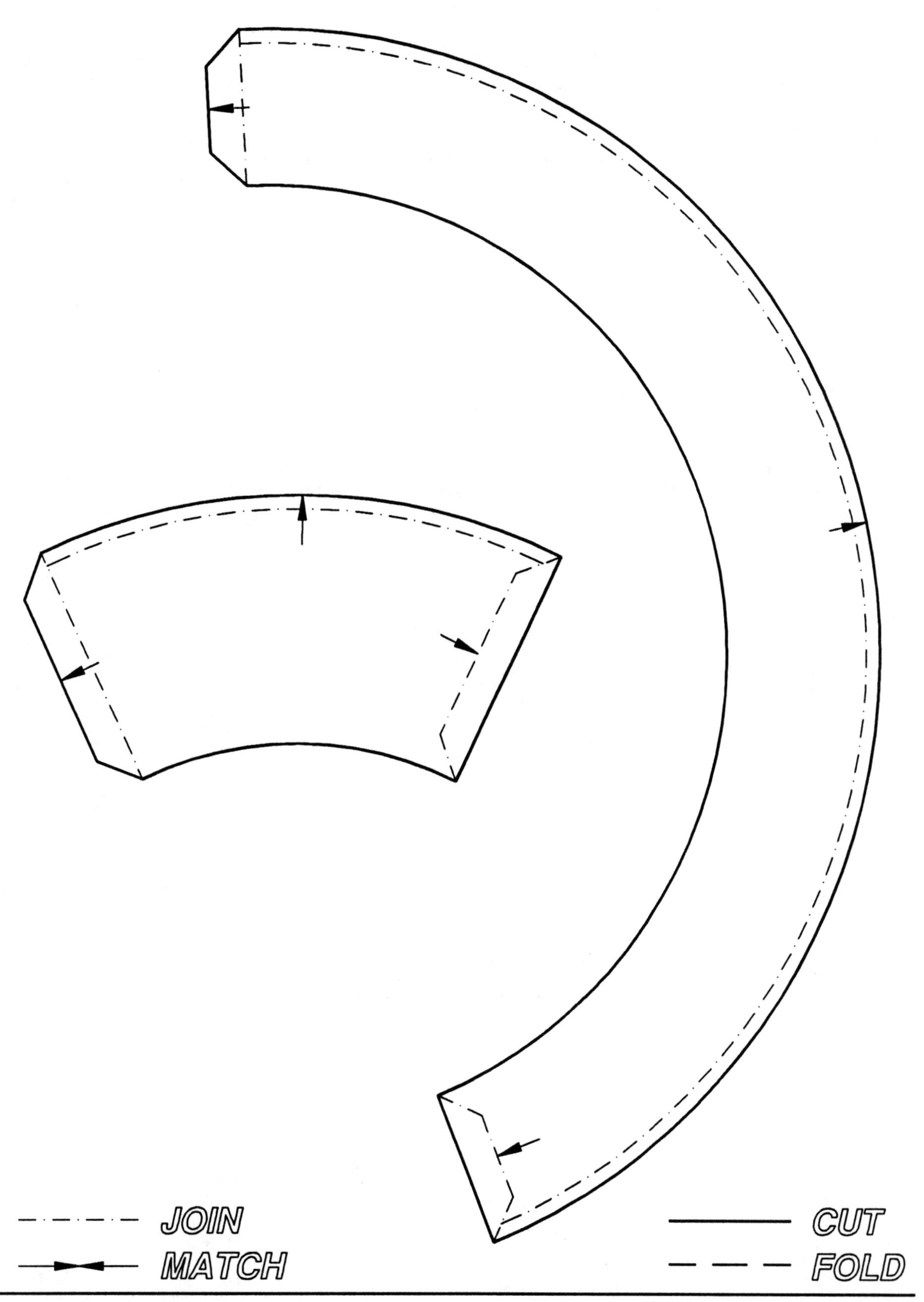

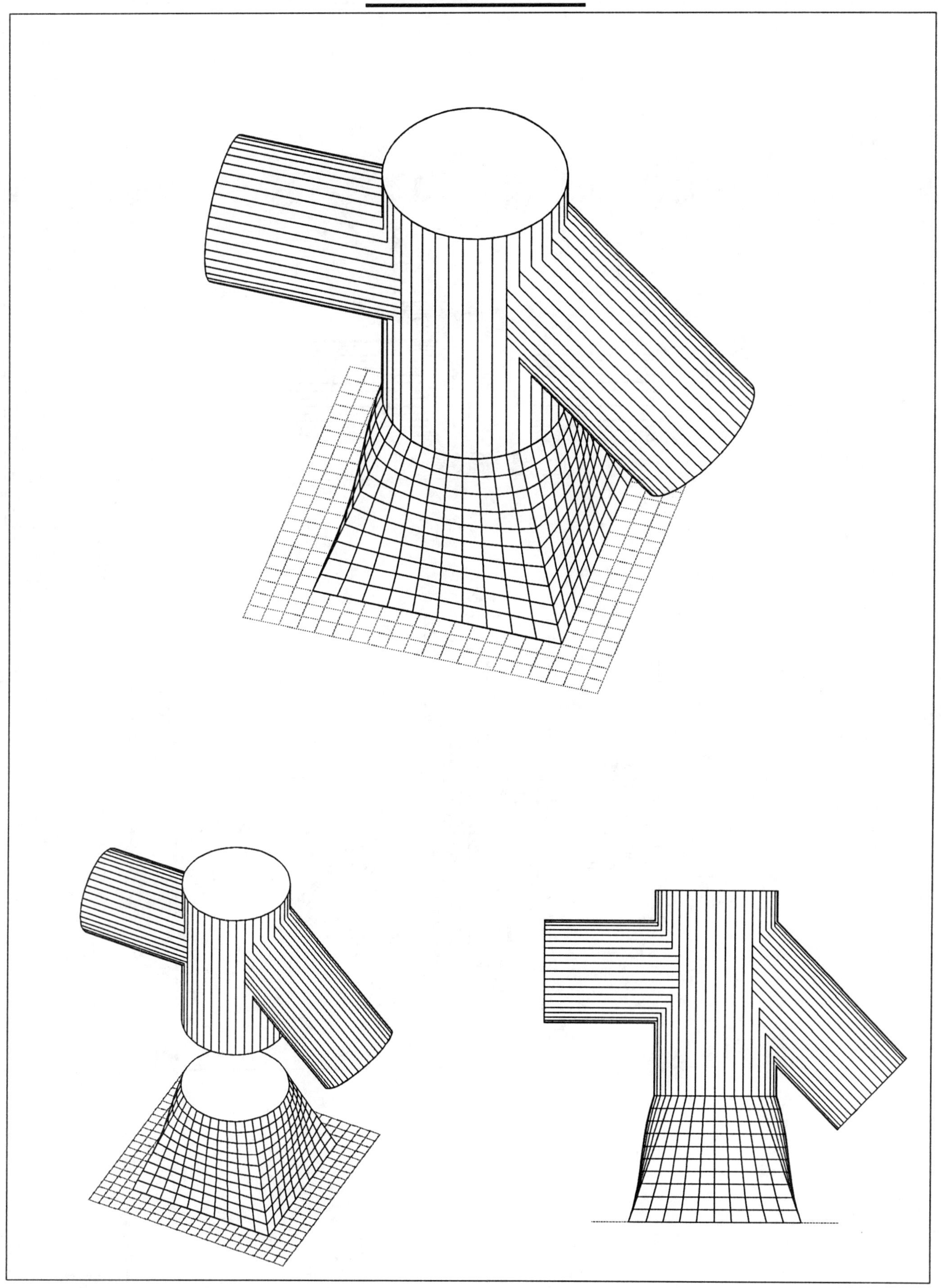

PROJECT-2

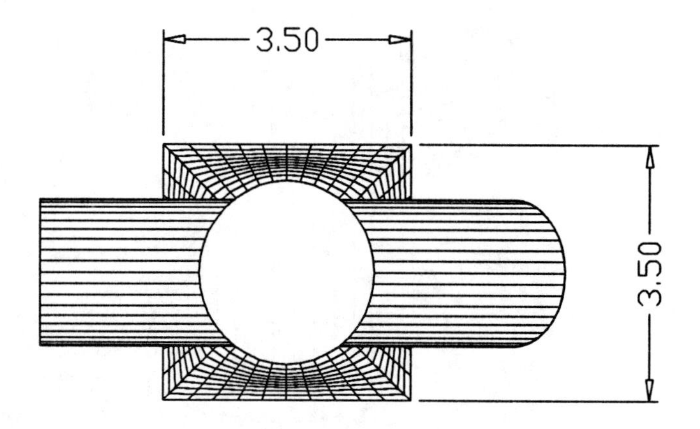

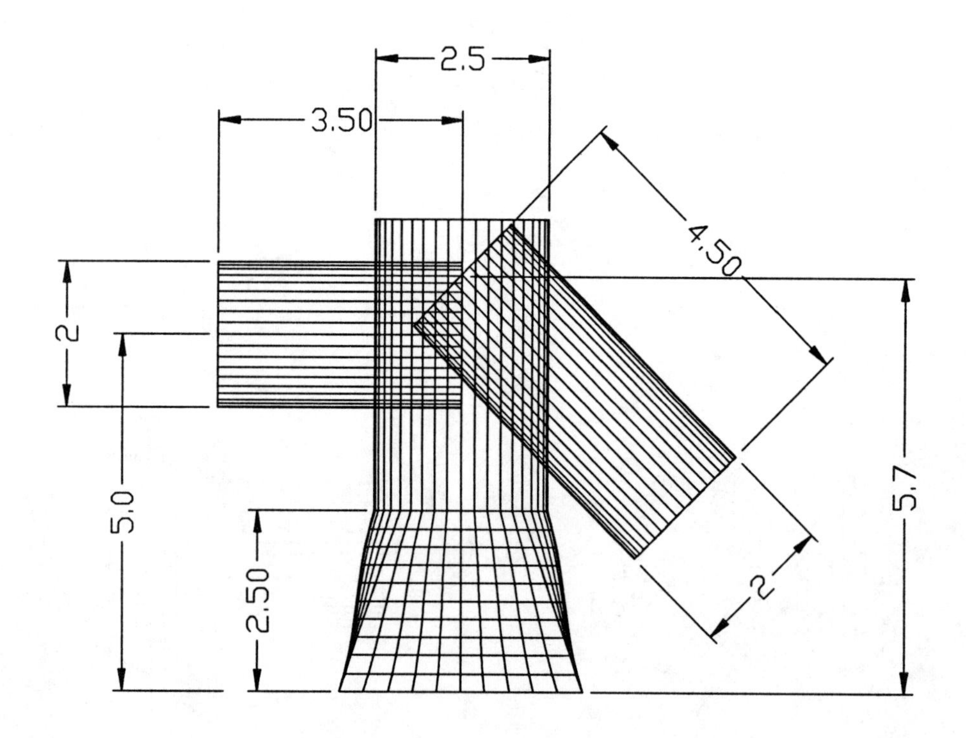

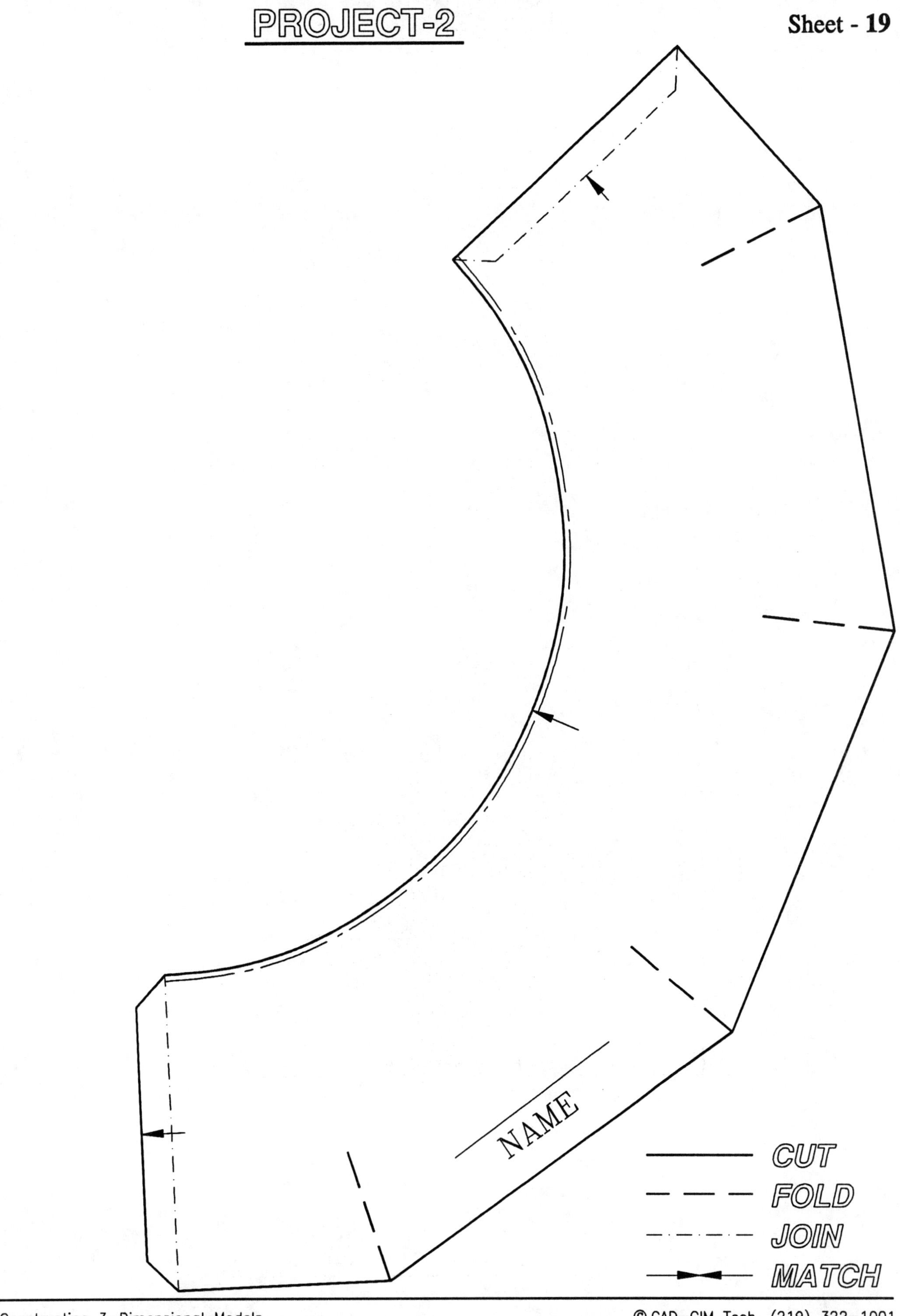
NAME
CUT
FOLD
JOIN
MATCH

PROJECT-2

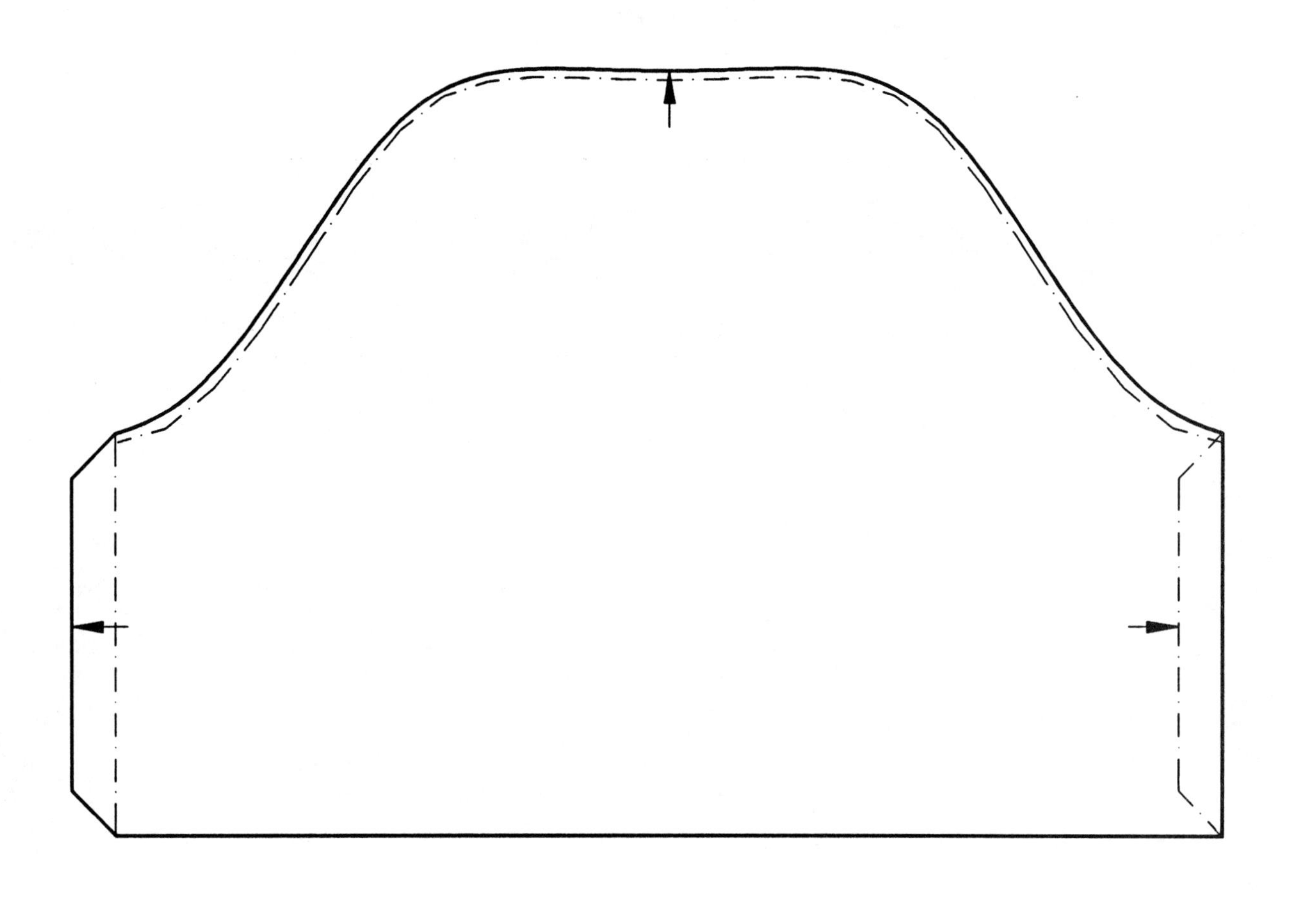

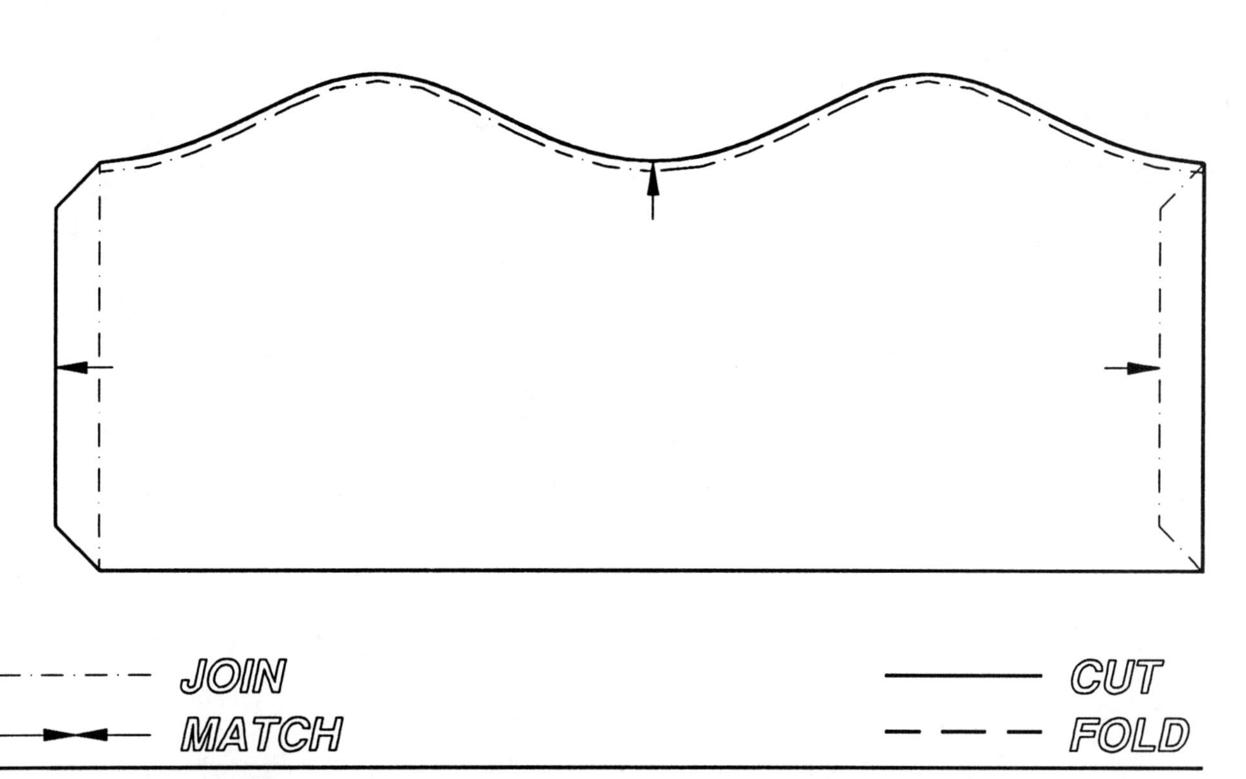

JOIN

MATCH

CUT

FOLD

PROJECT-2

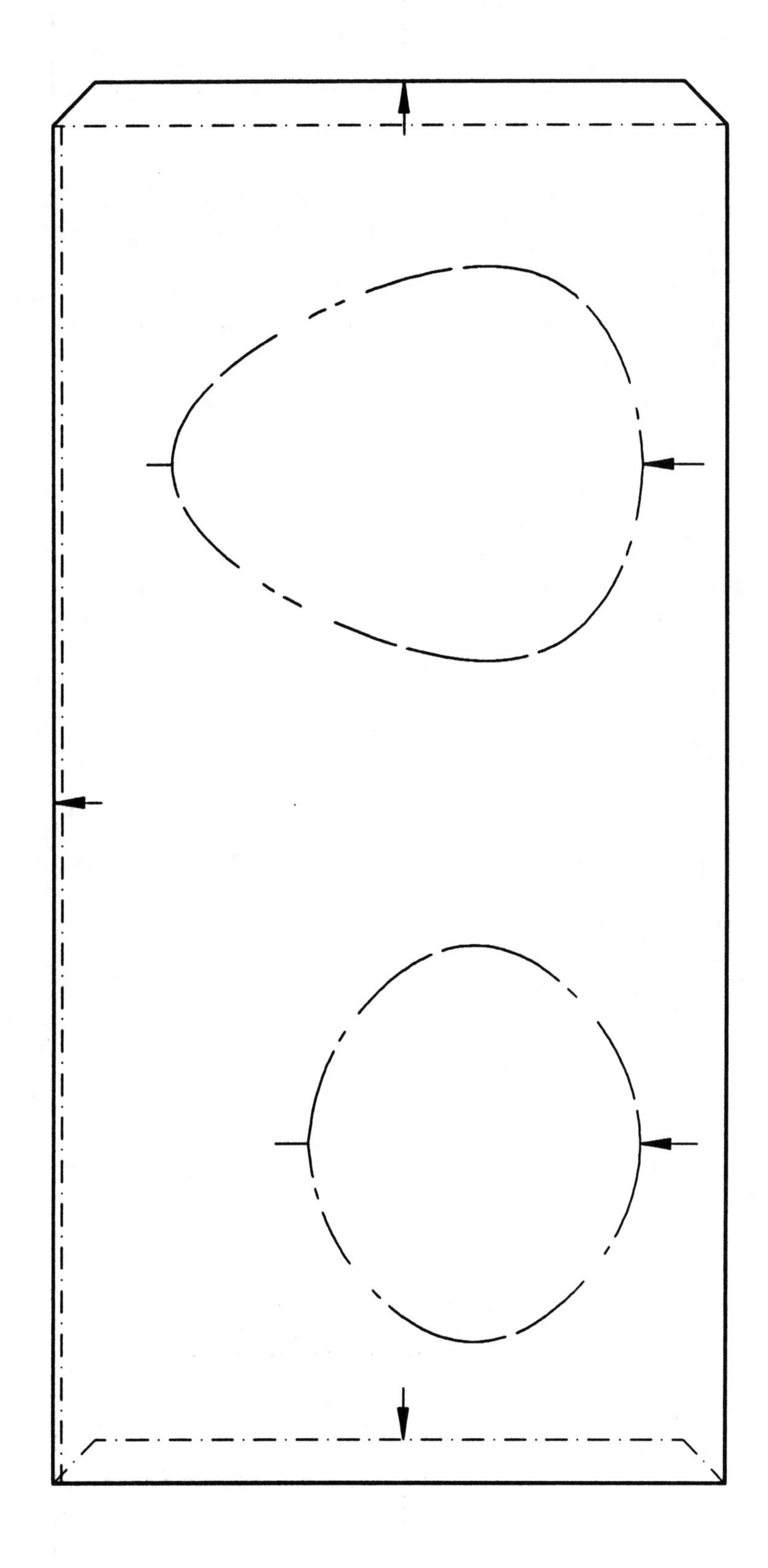

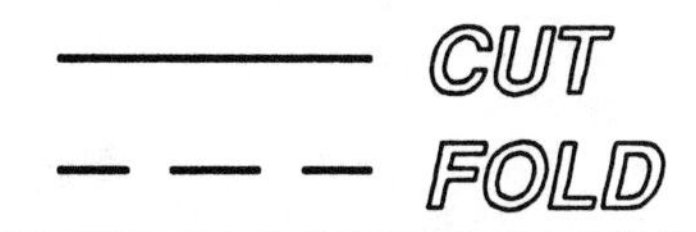

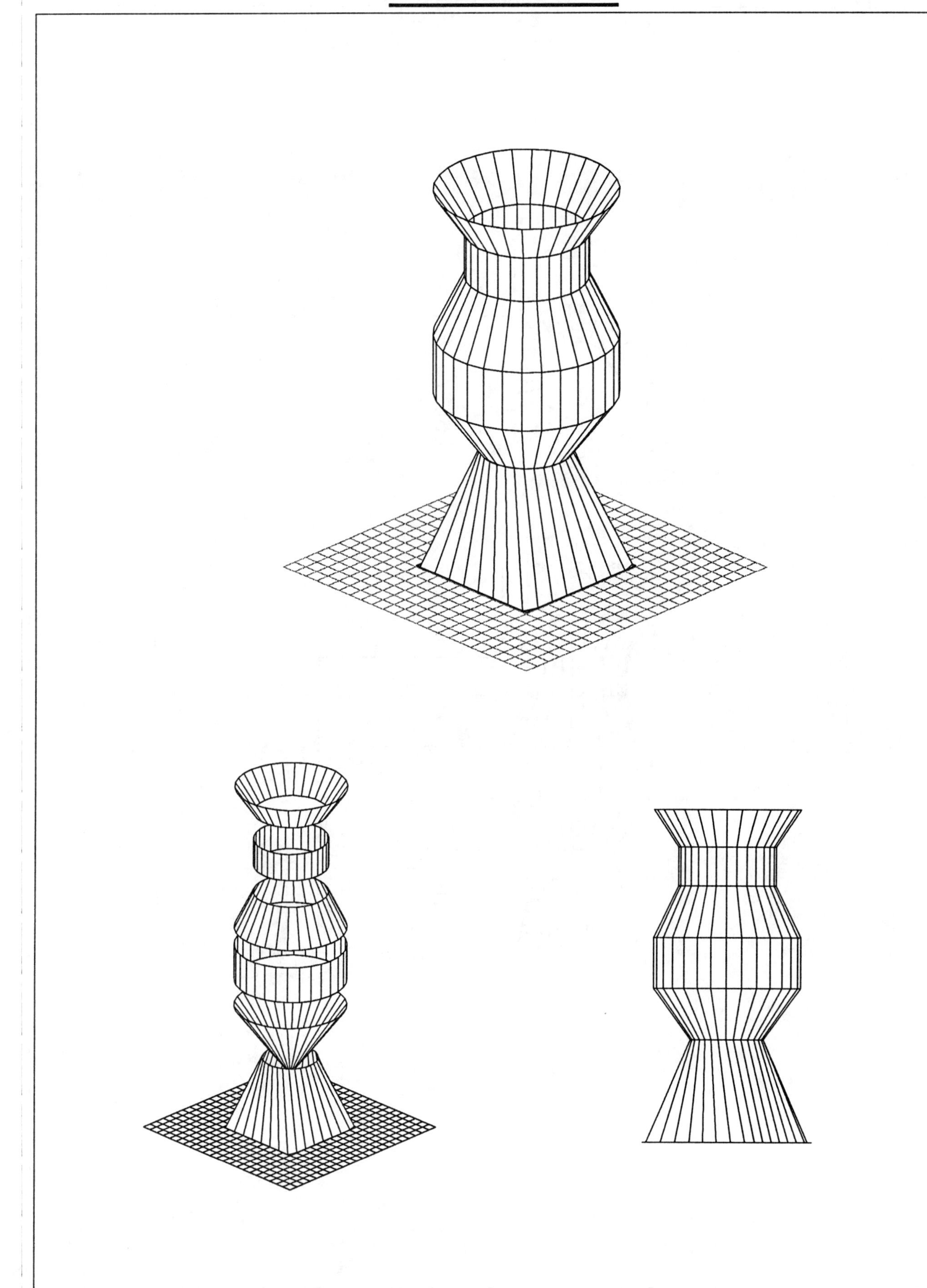

PROJECT-3

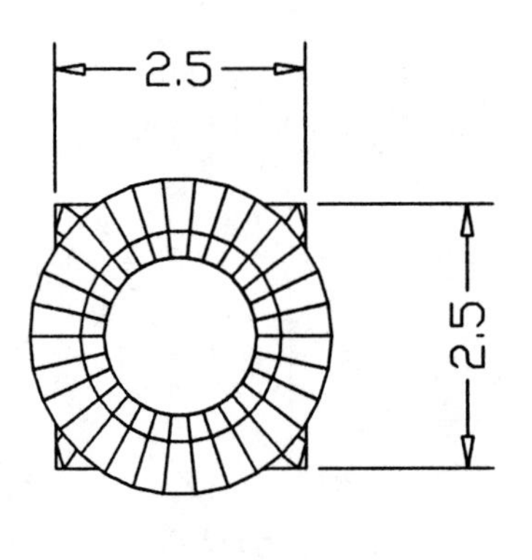

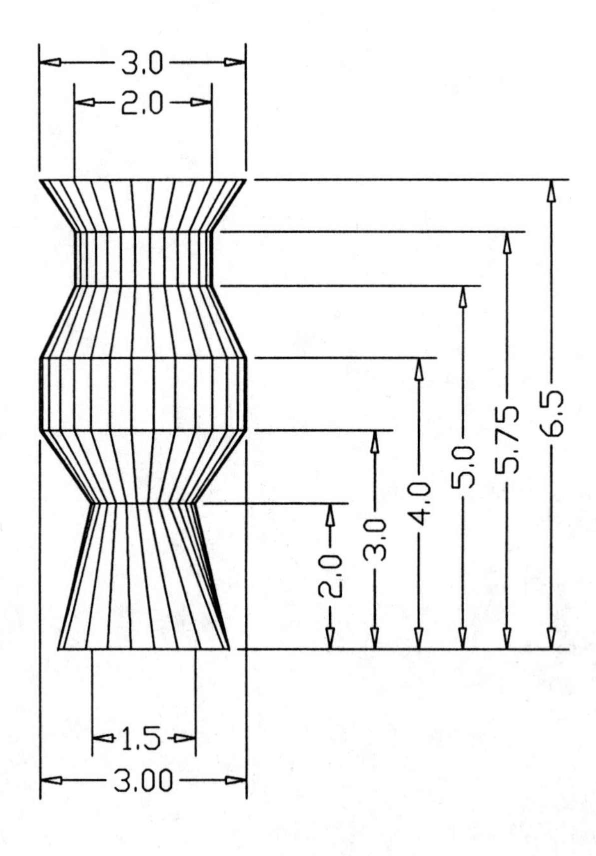

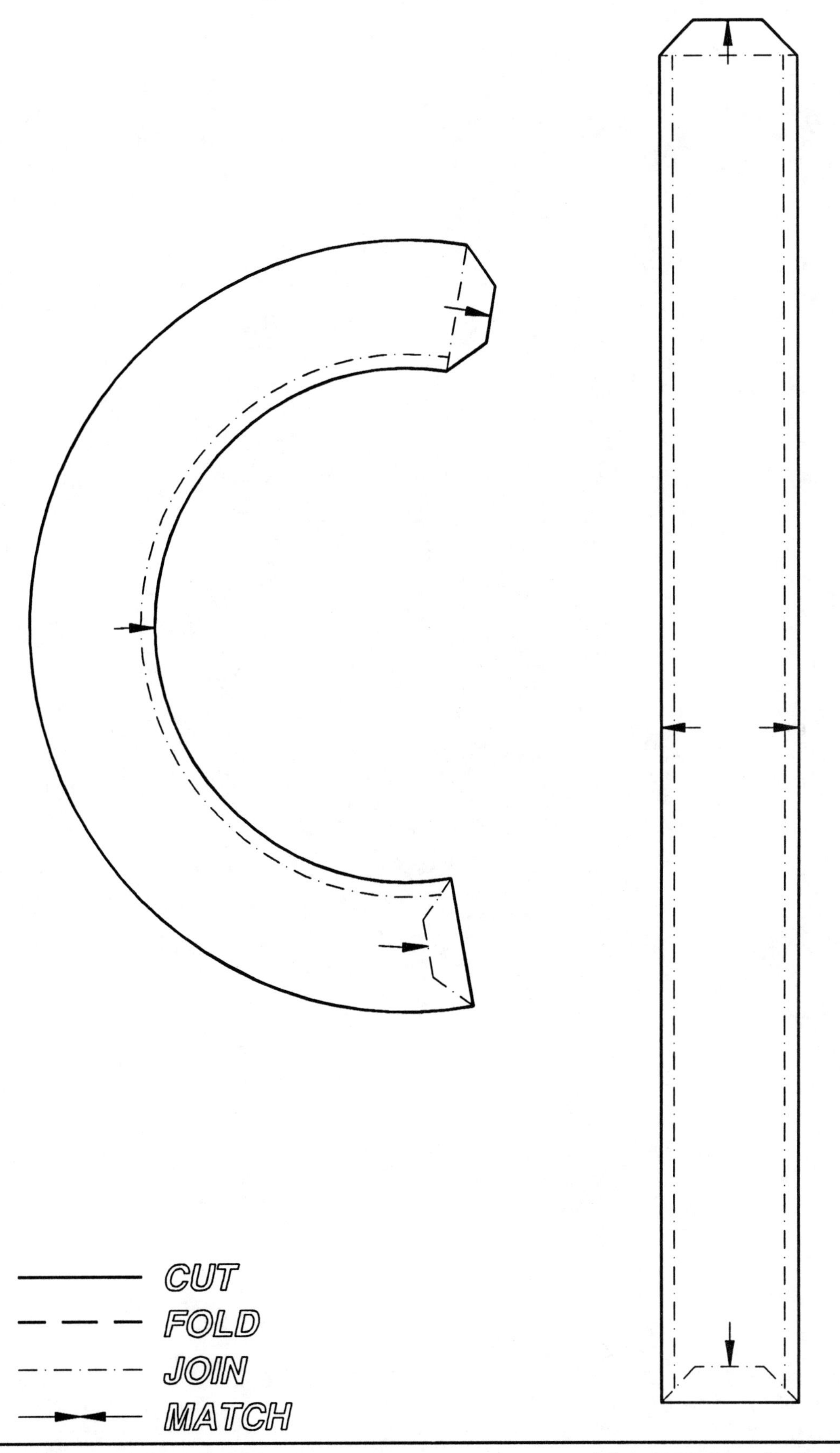
PROJECT-3
CUT
FOLD
JOIN
MATCH

PROJECT-3

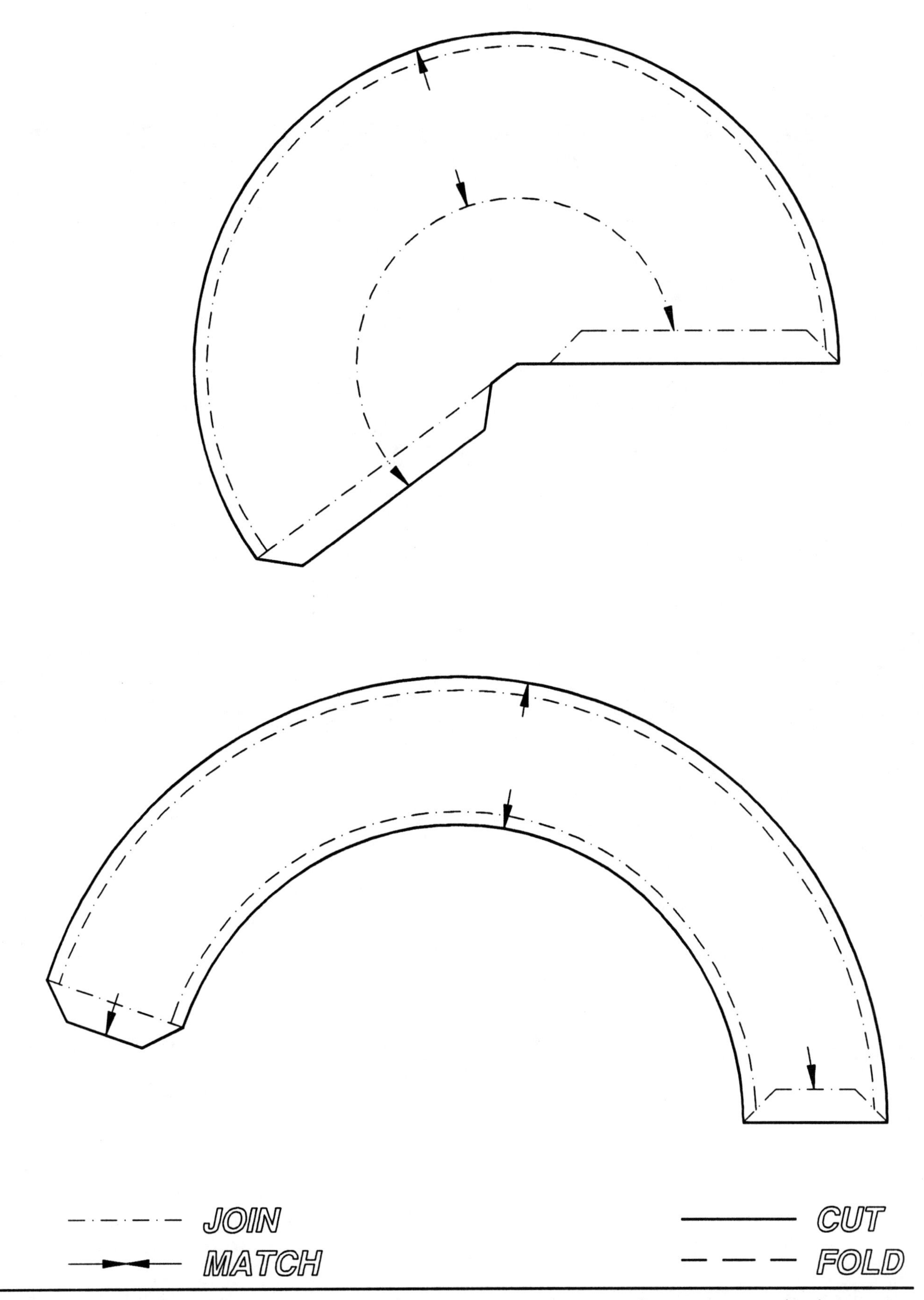

PROJECT-3

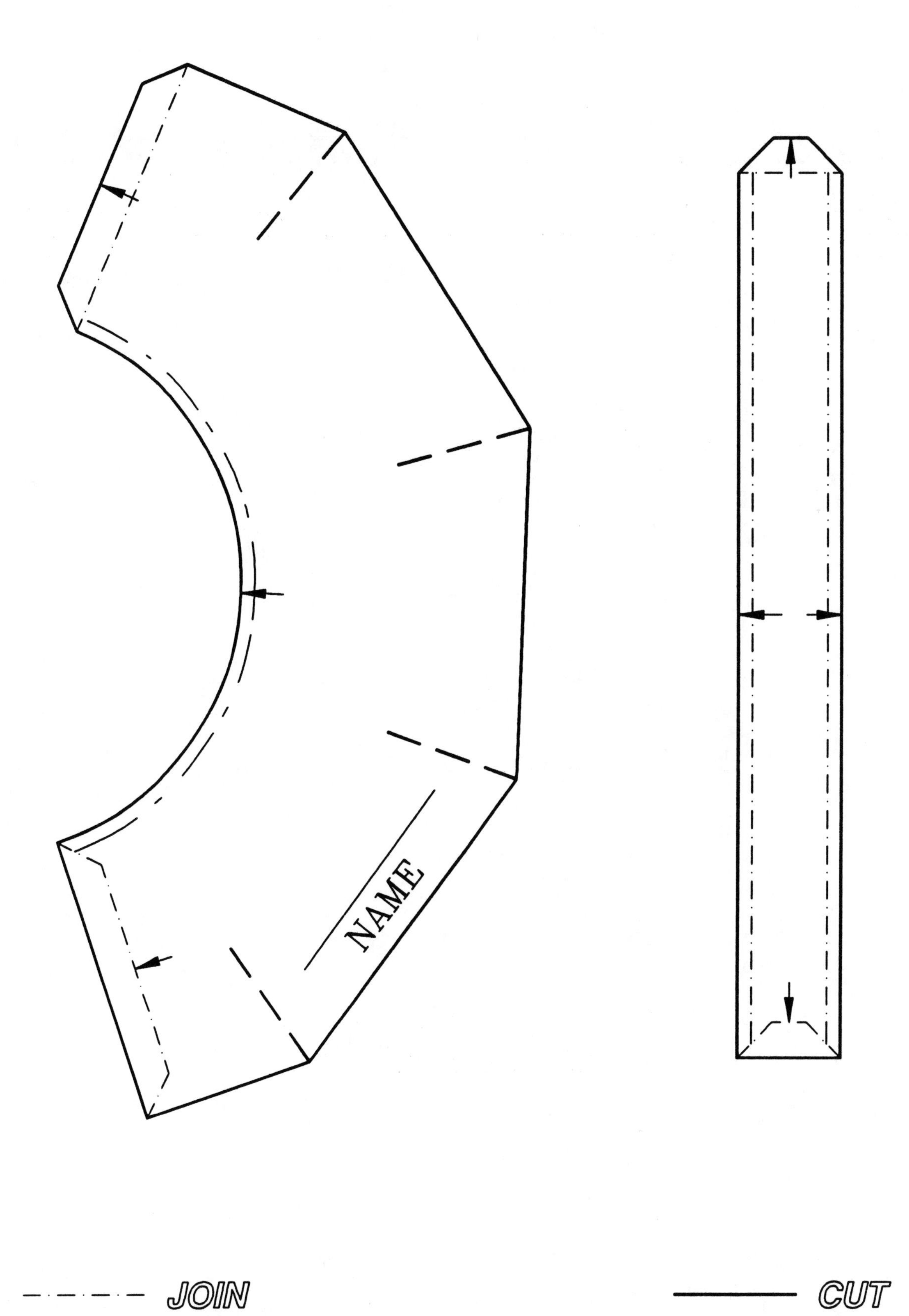

JOIN

MATCH

CUT

FOLD

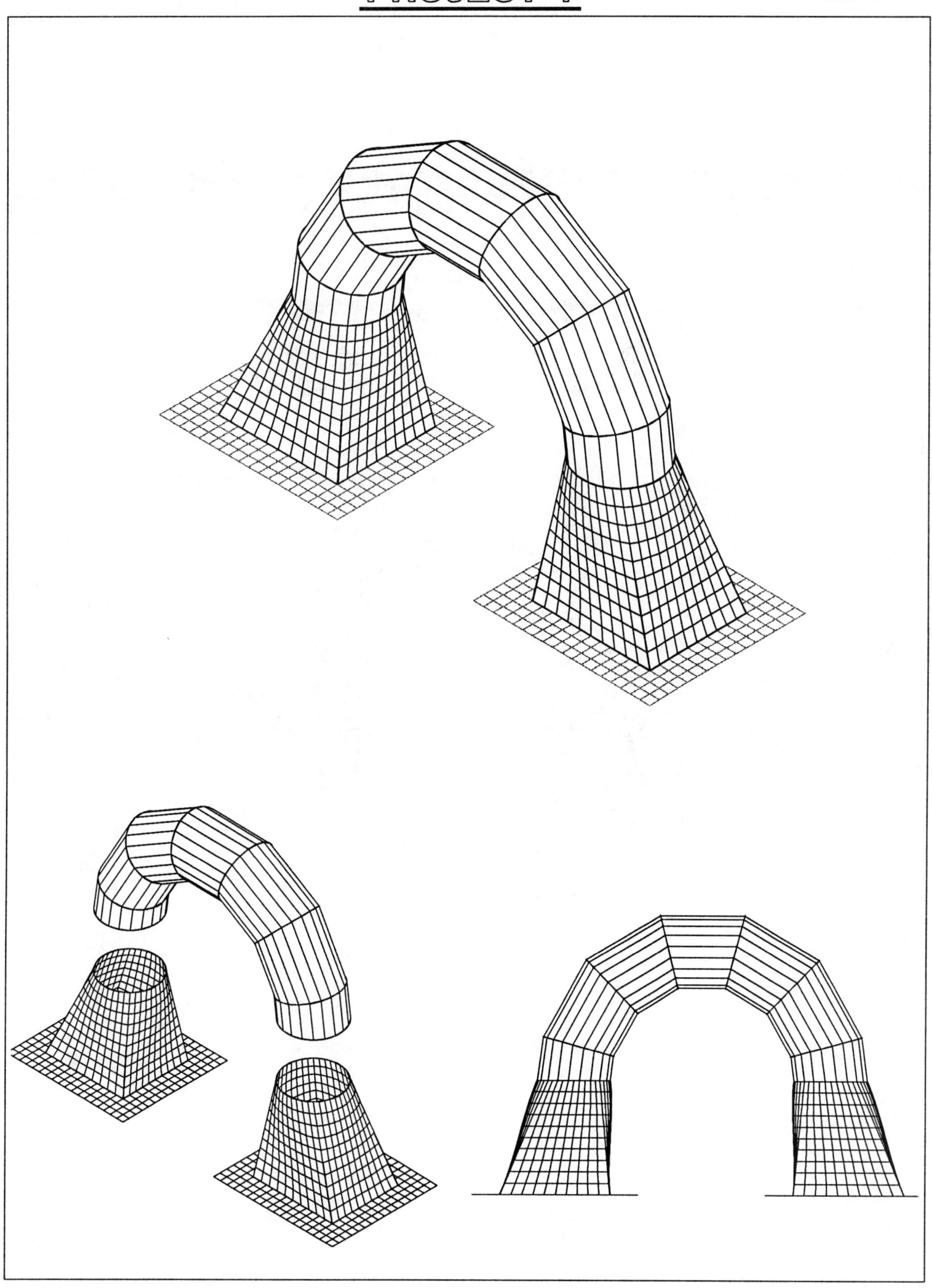

PROJECT-4

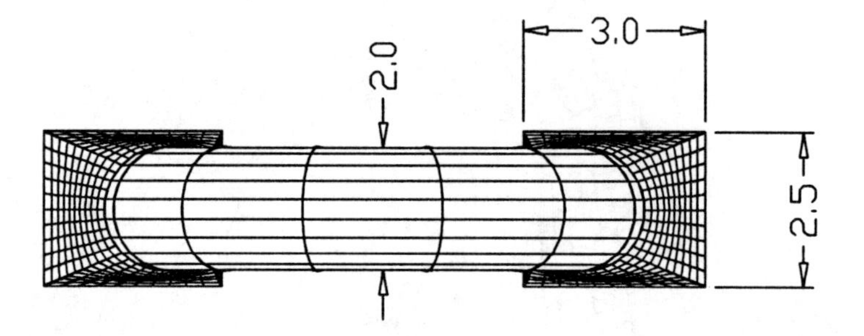

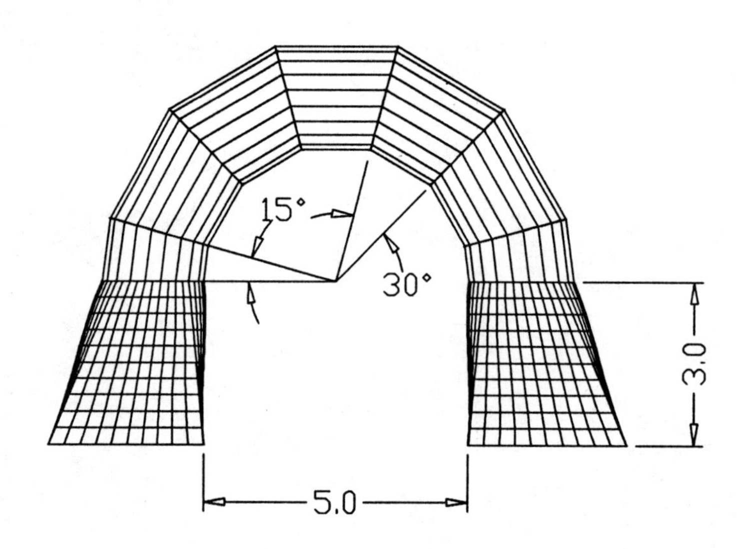

PROJECT-4

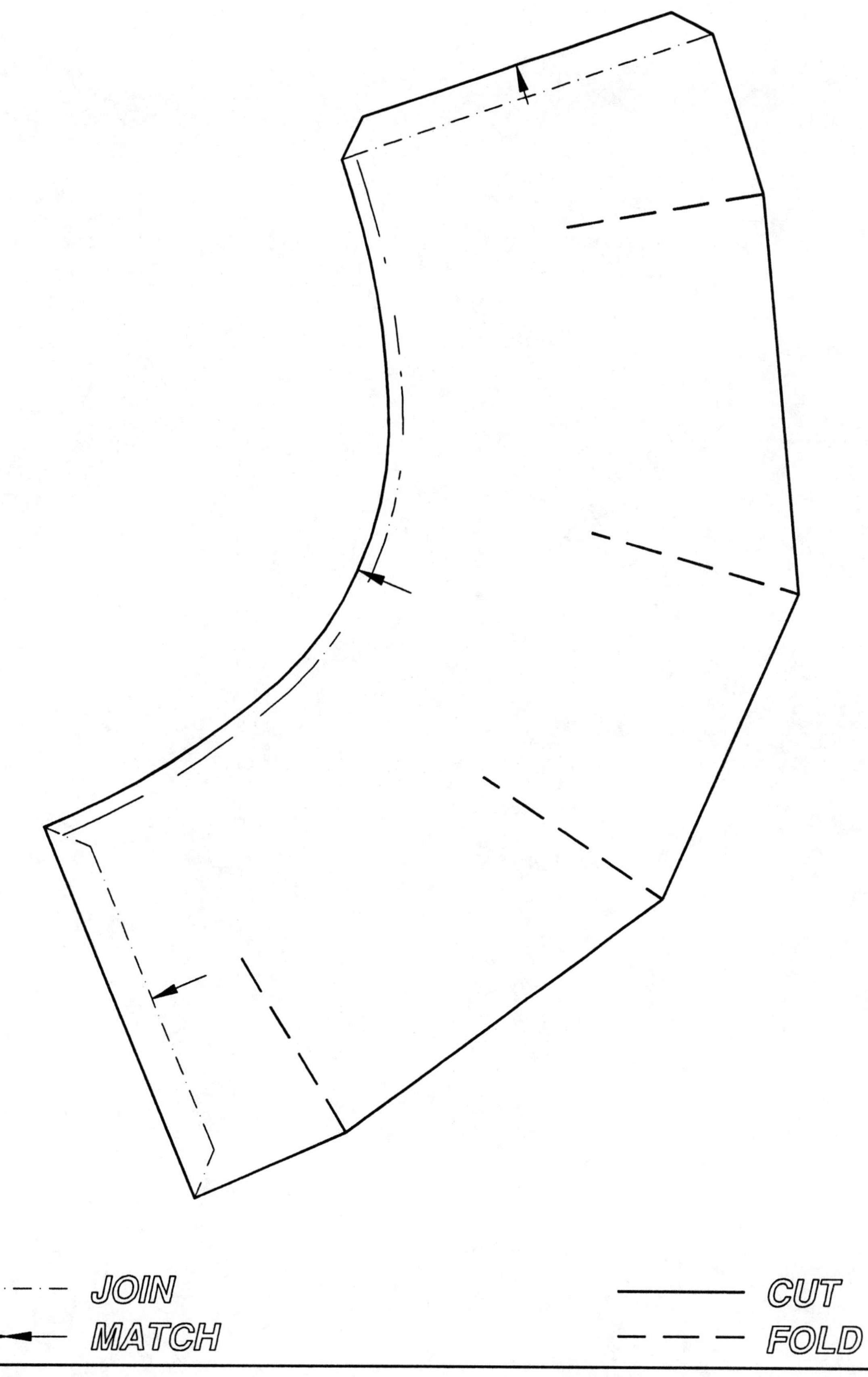

PROJECT-4

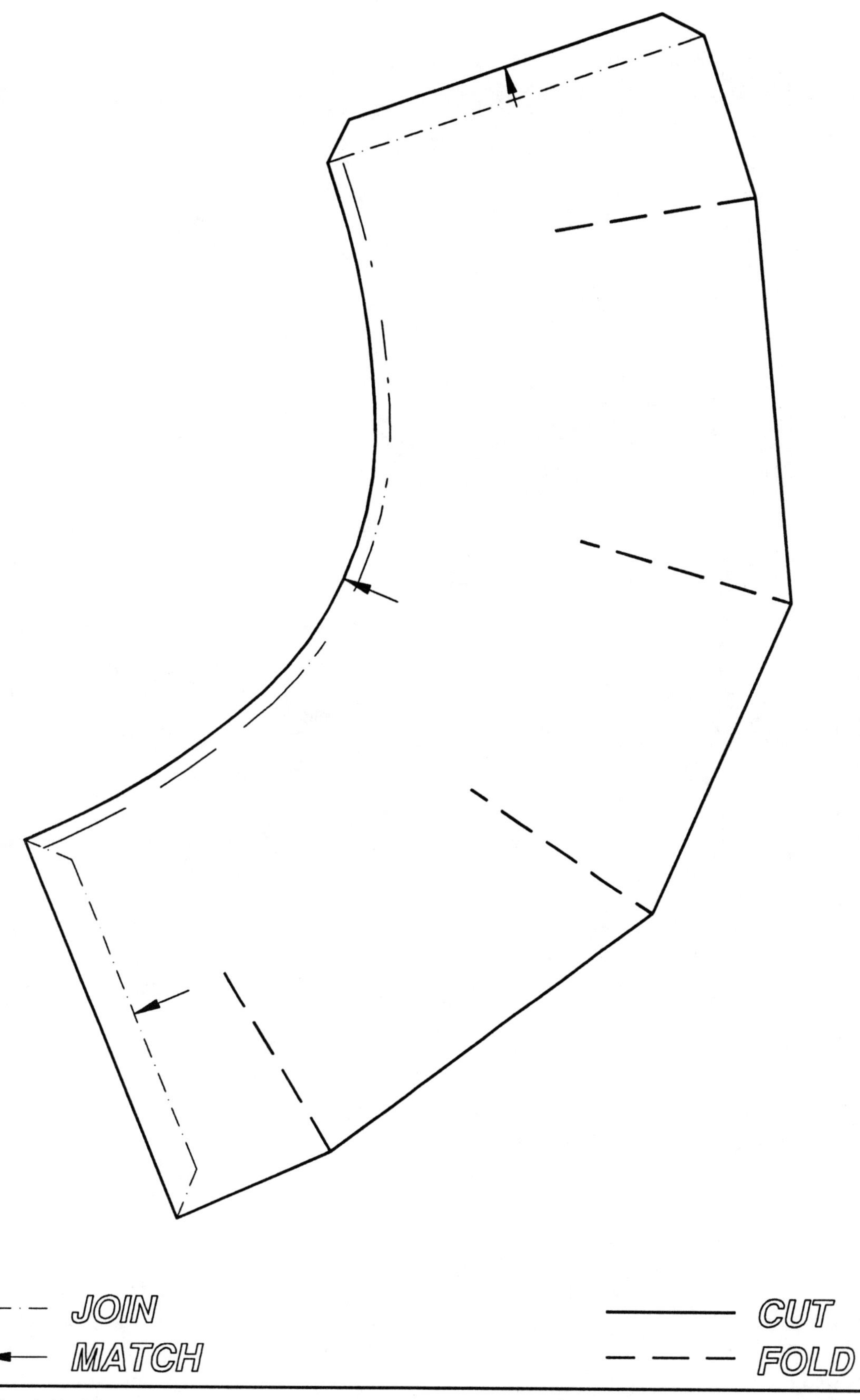

JOIN

MATCH

CUT

FOLD

PROJECT-4

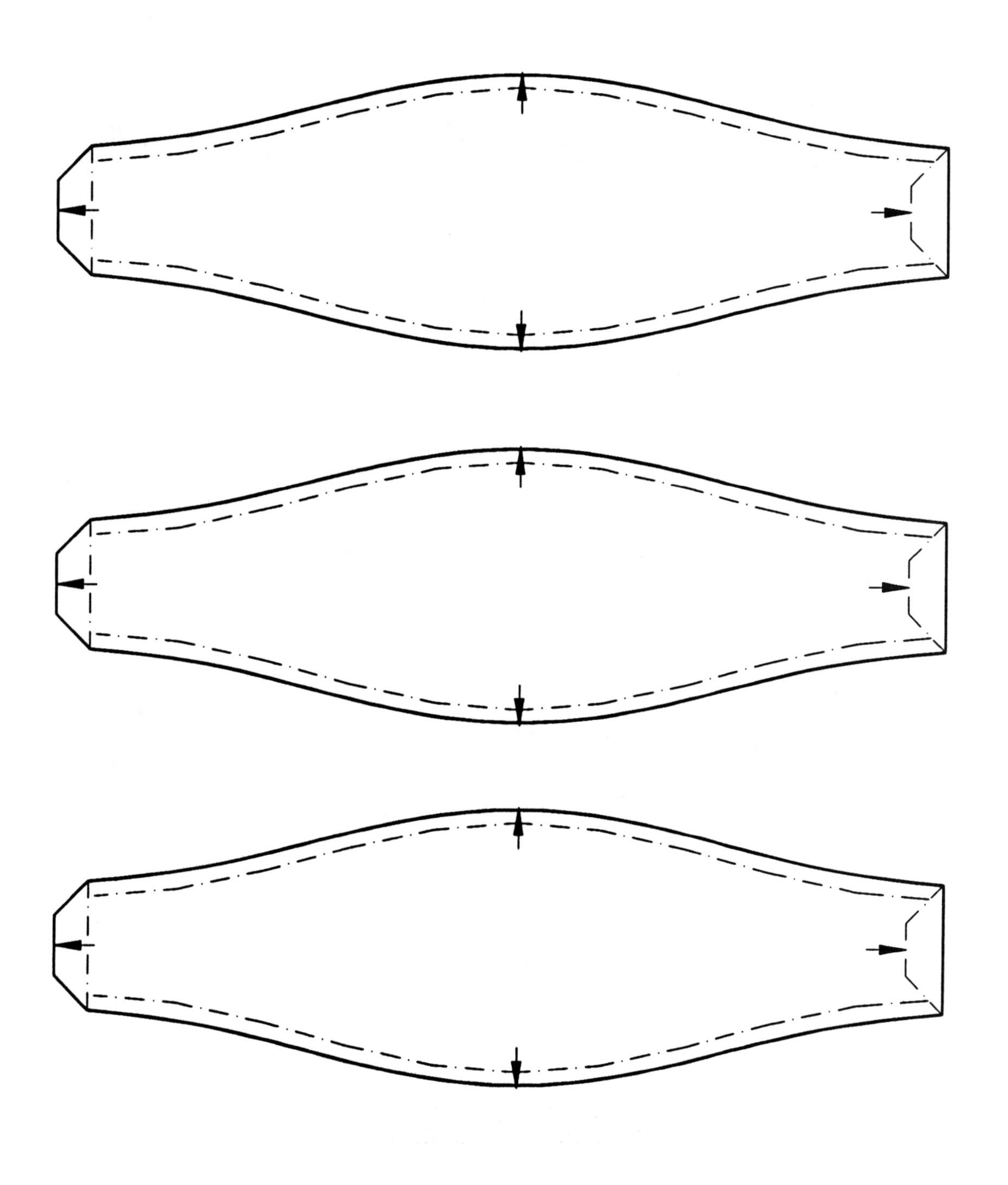

CUT

FOLD

PROJECT-4

Sheet - 30

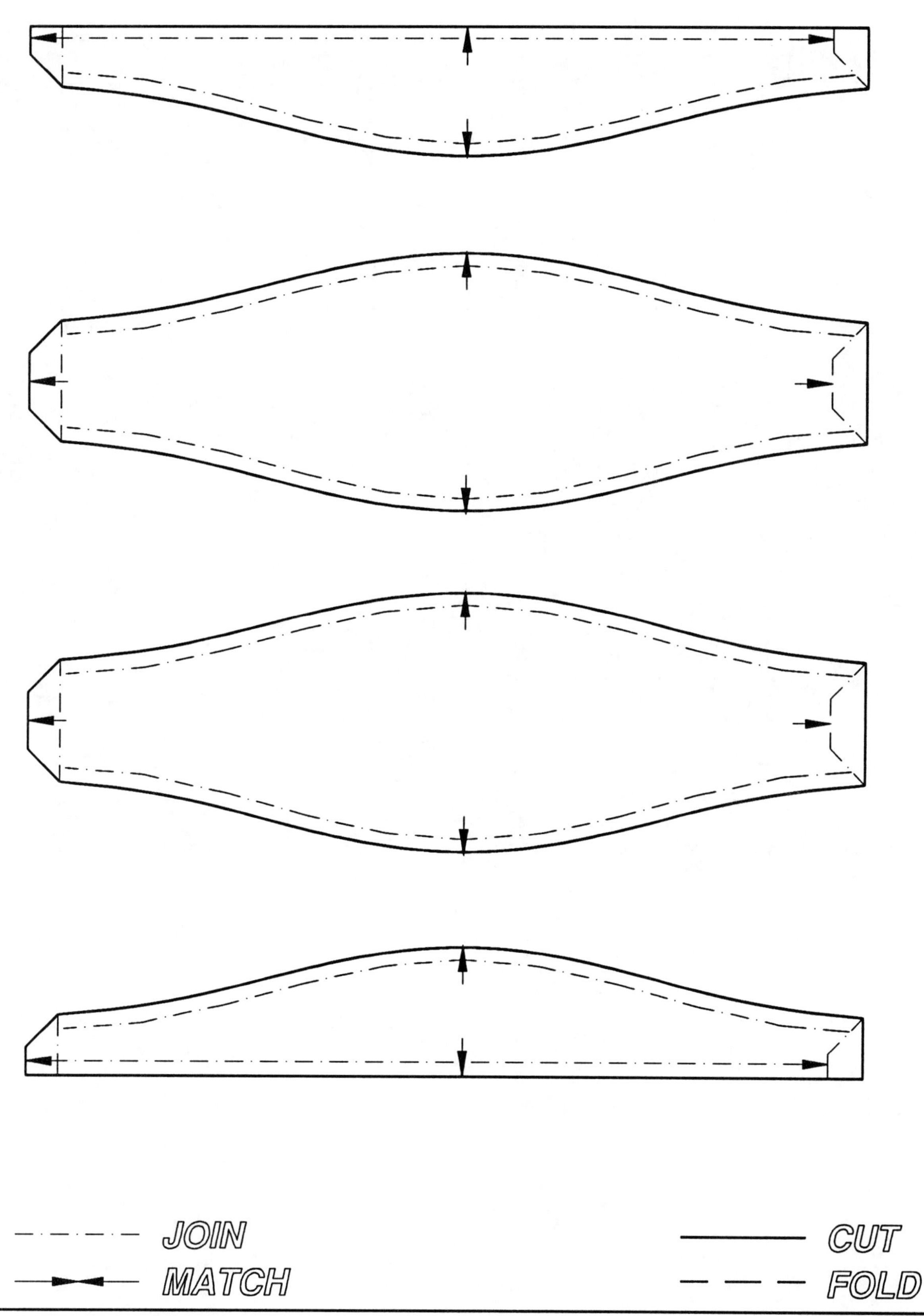

PROJECT-5

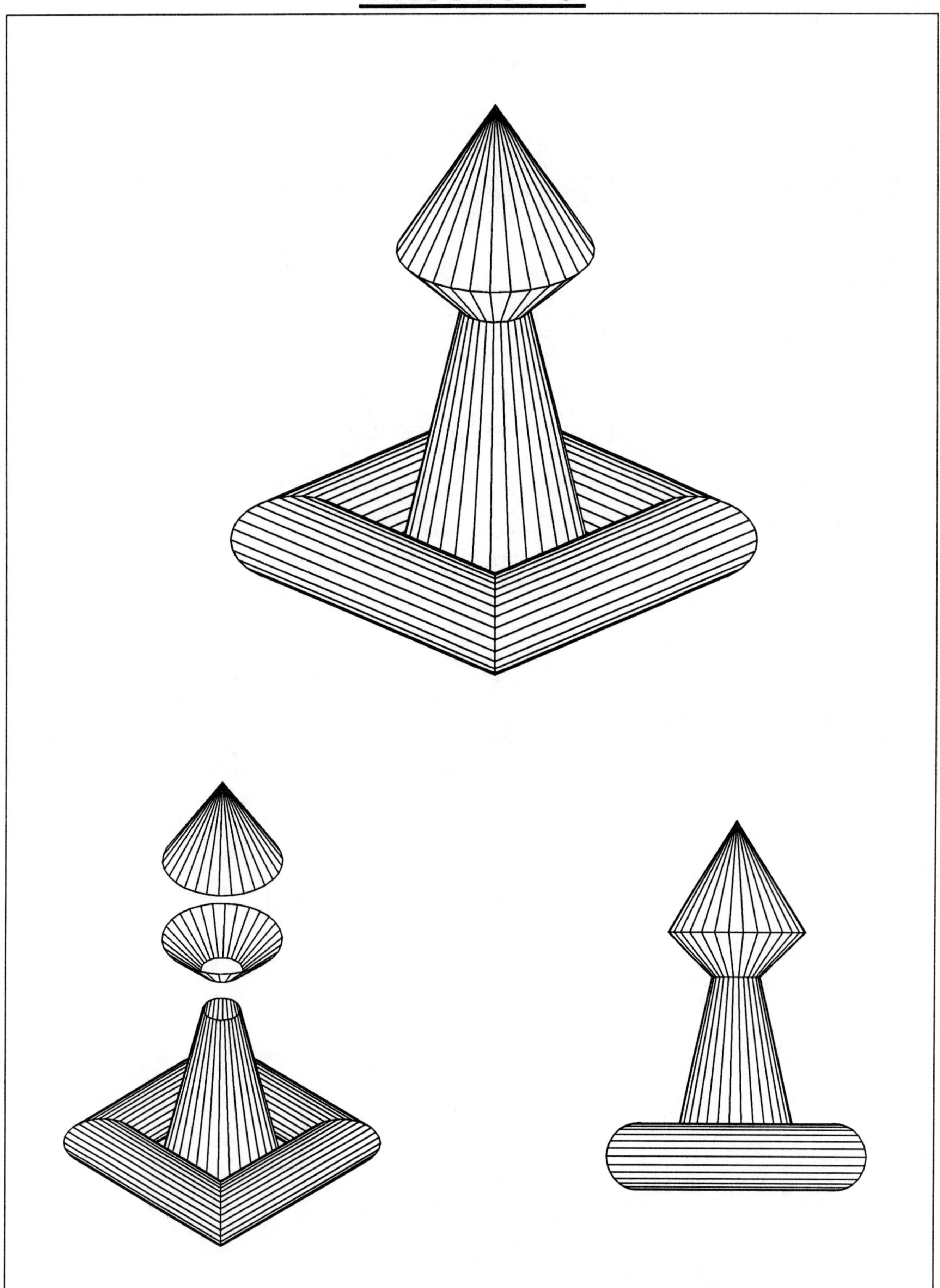

PROJECT-5

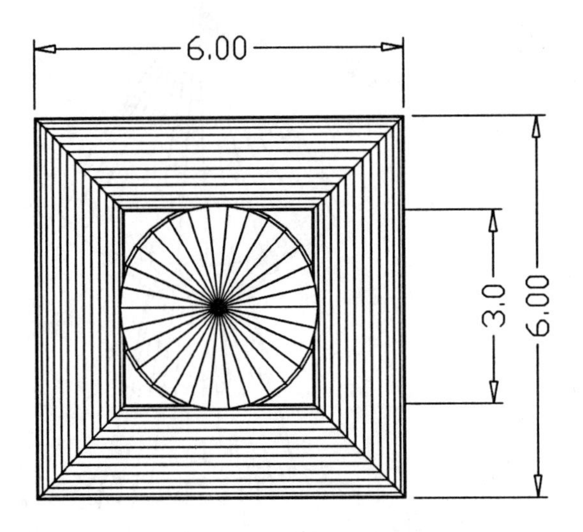
6.00
3.0
6.00

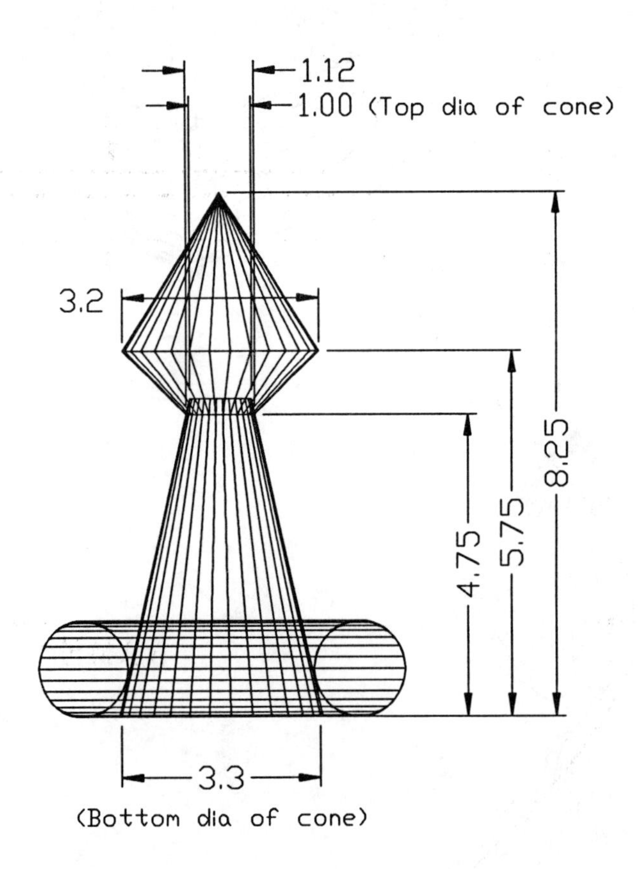
1.12
1.00 (Top dia of cone)
3.2
8.25
5.75
4.75
3.3
(Bottom dia of cone)

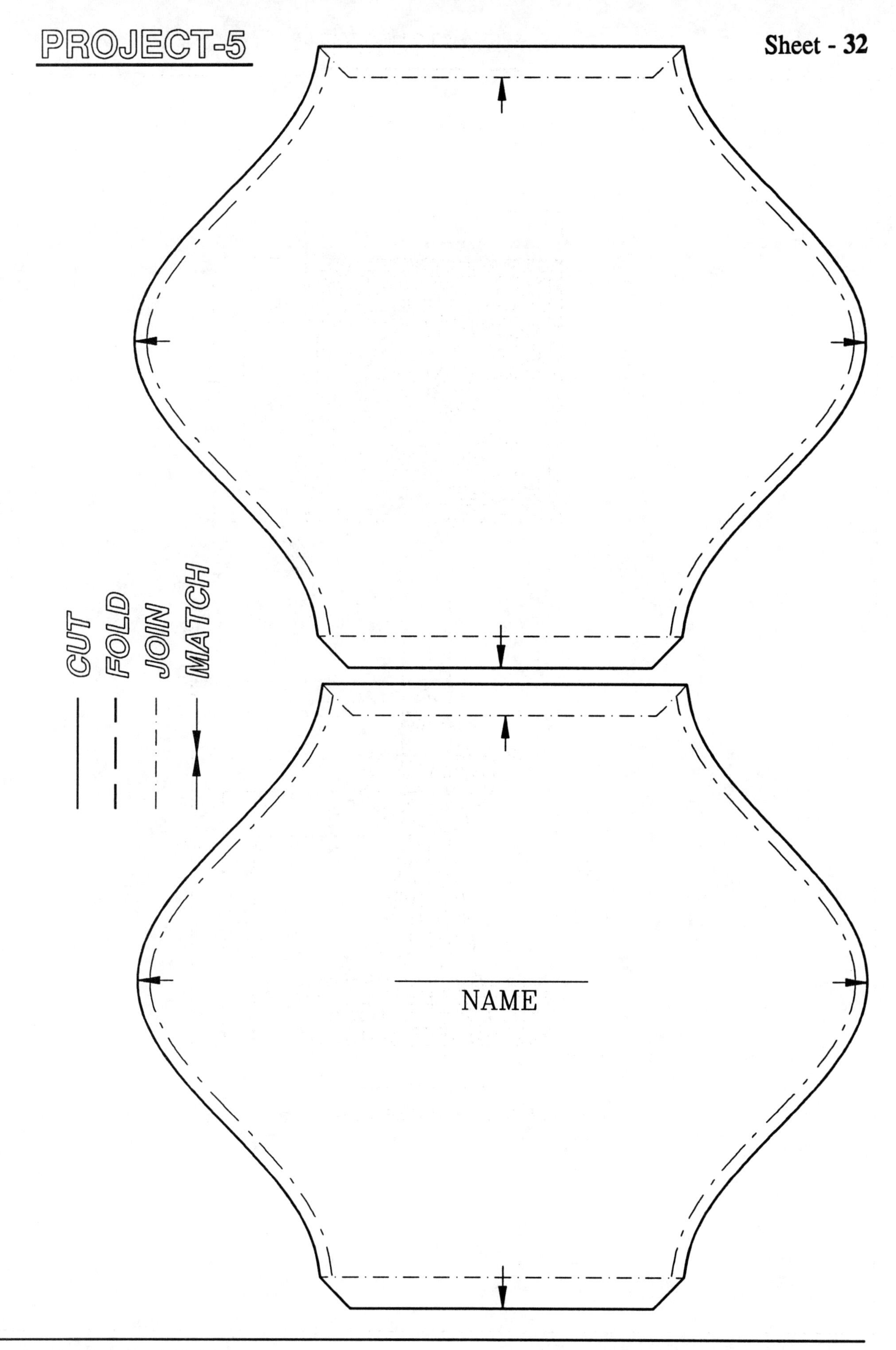
CUT
FOLD
JOIN
MATCH
NAME

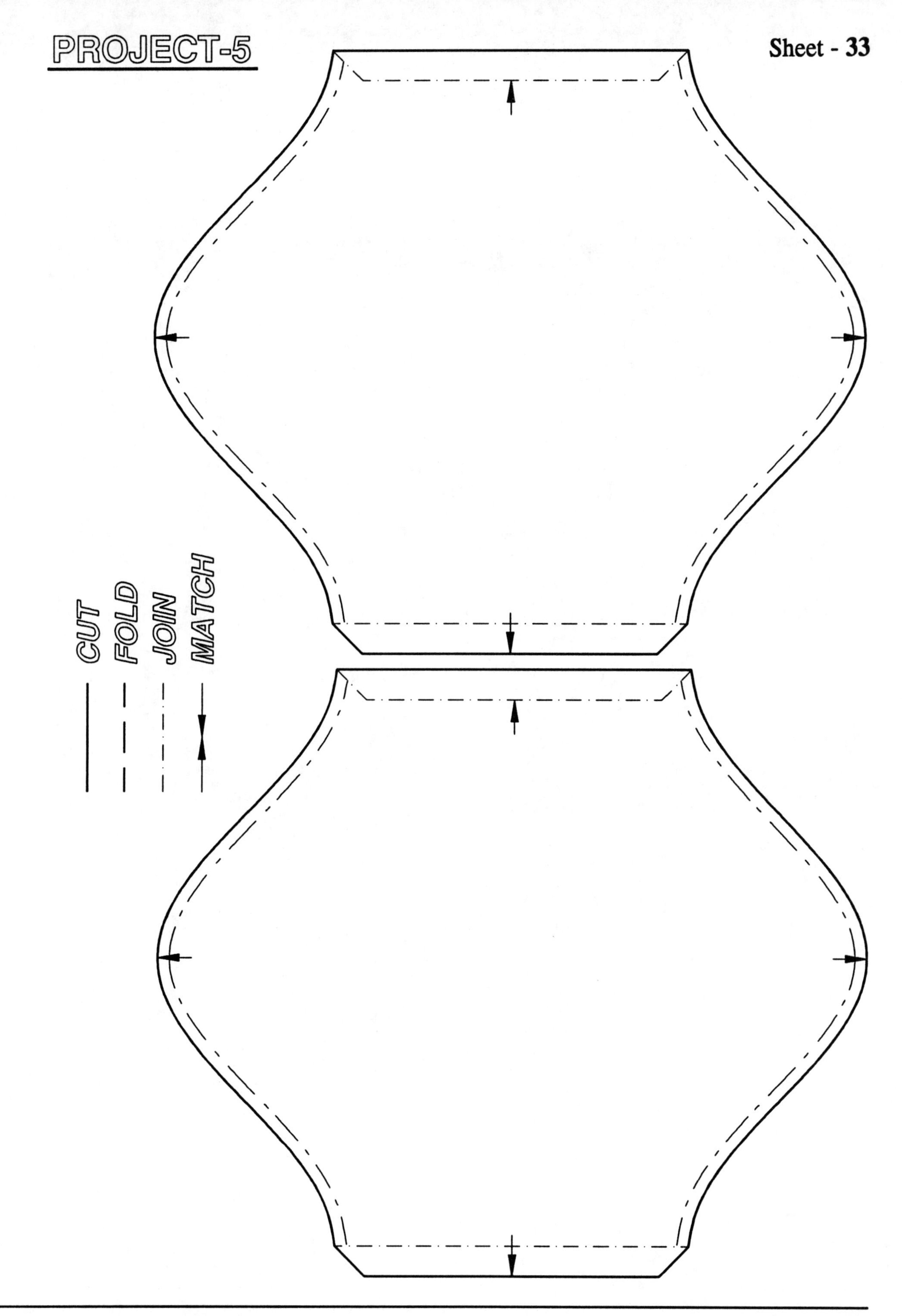
CUT
FOLD
JOIN
MATCH

PROJECT-5

Sheet - **34**

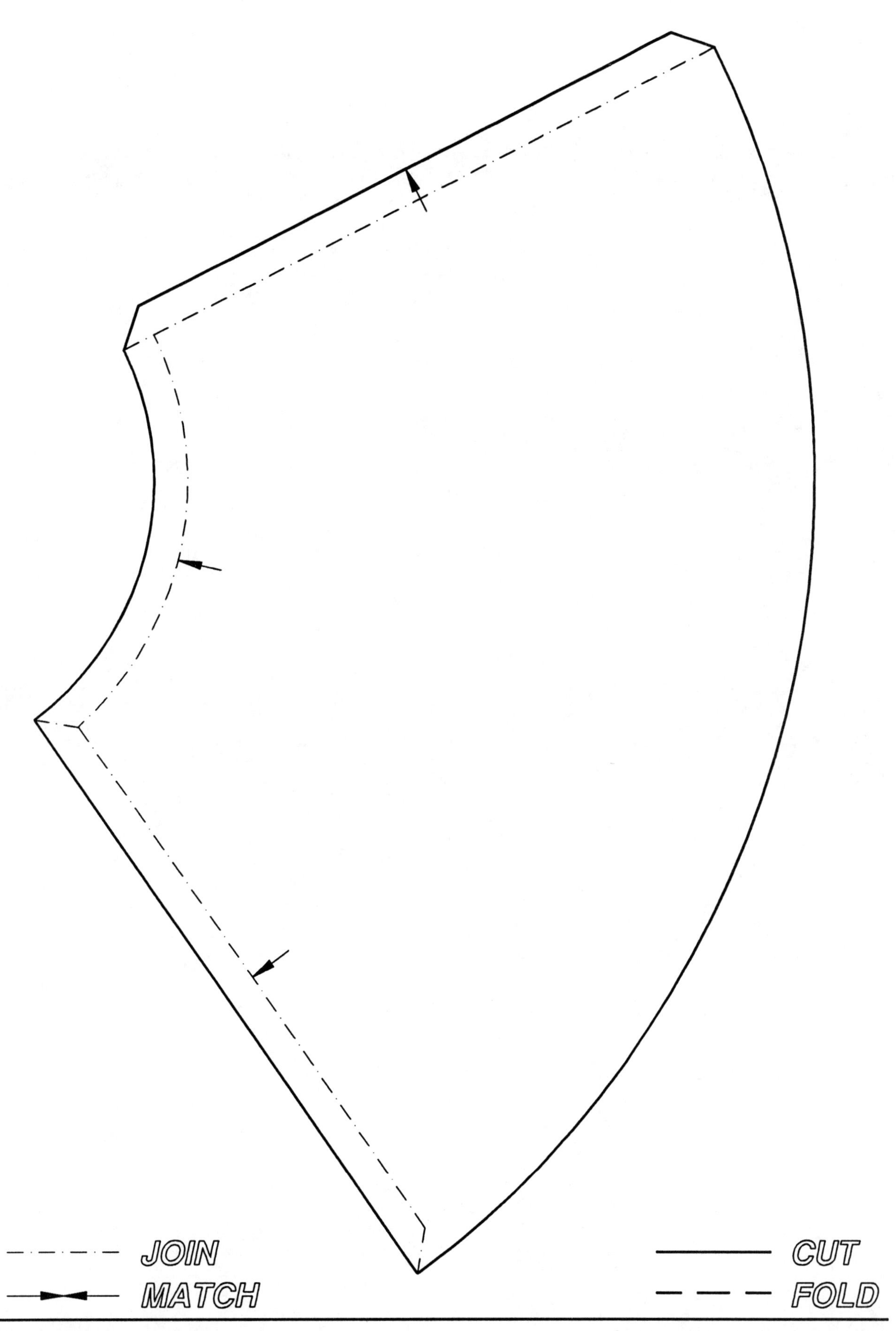

PROJECT-5

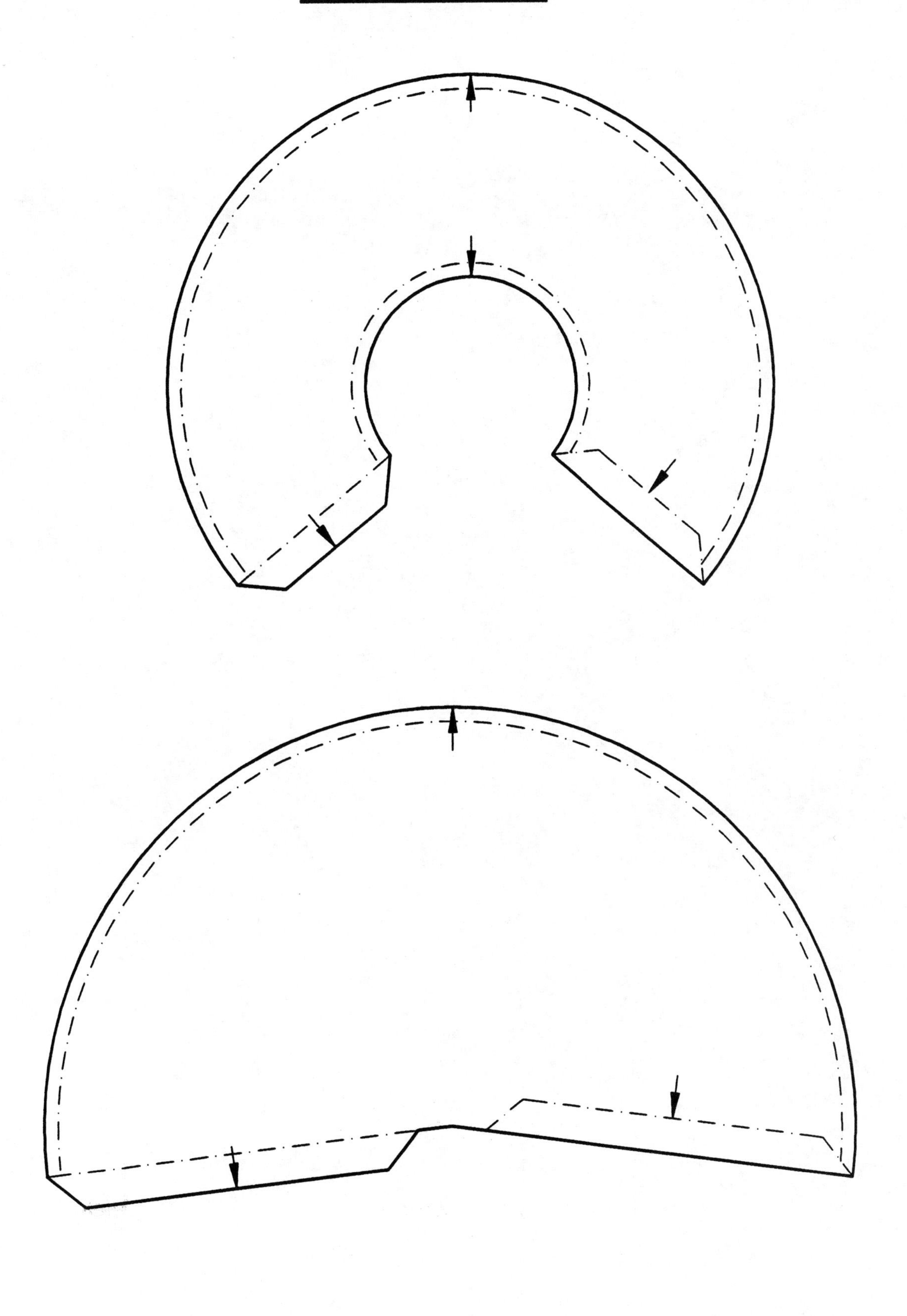

CUT
FOLD

PROJECT-6

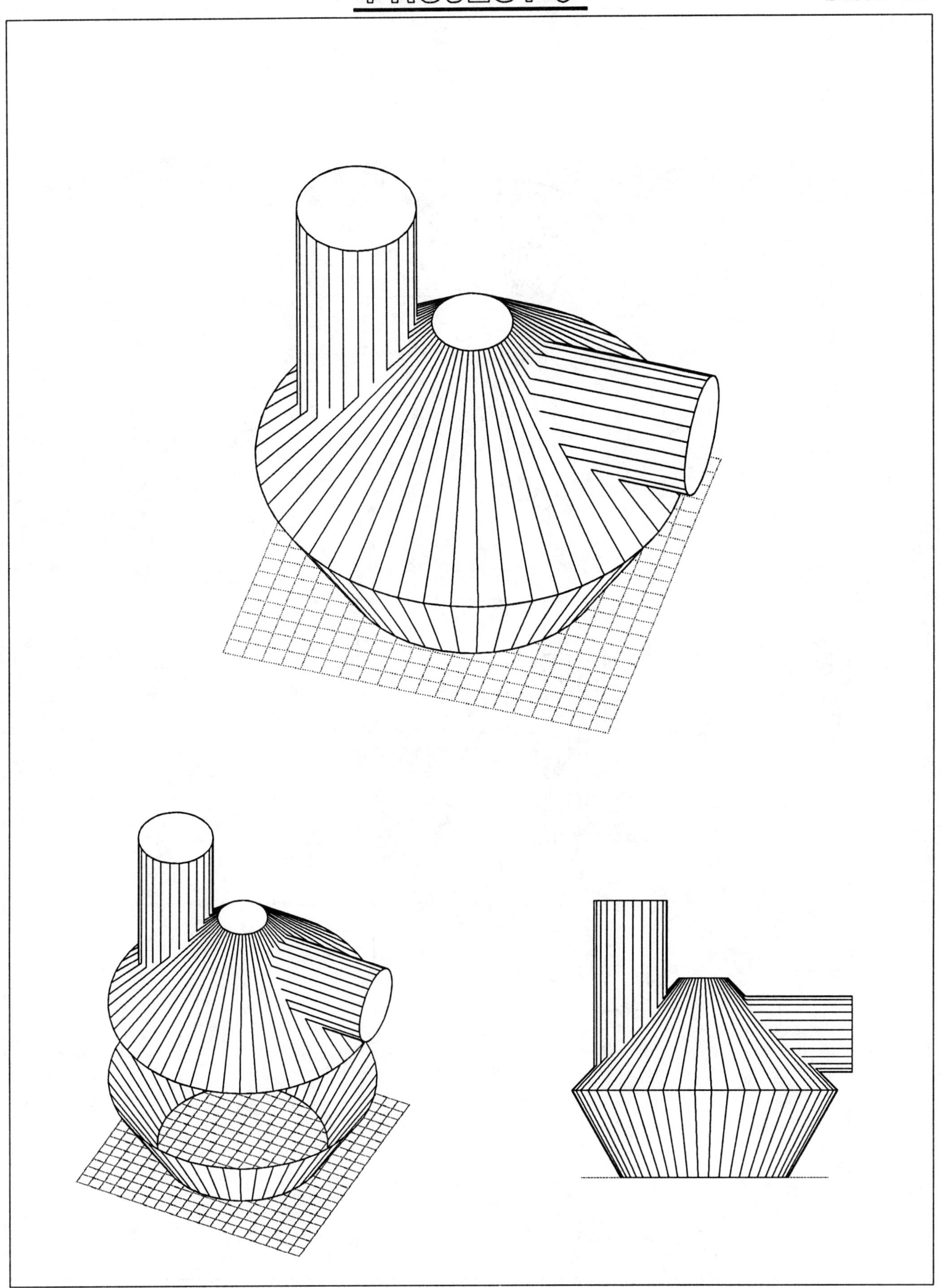

PROJECT-6

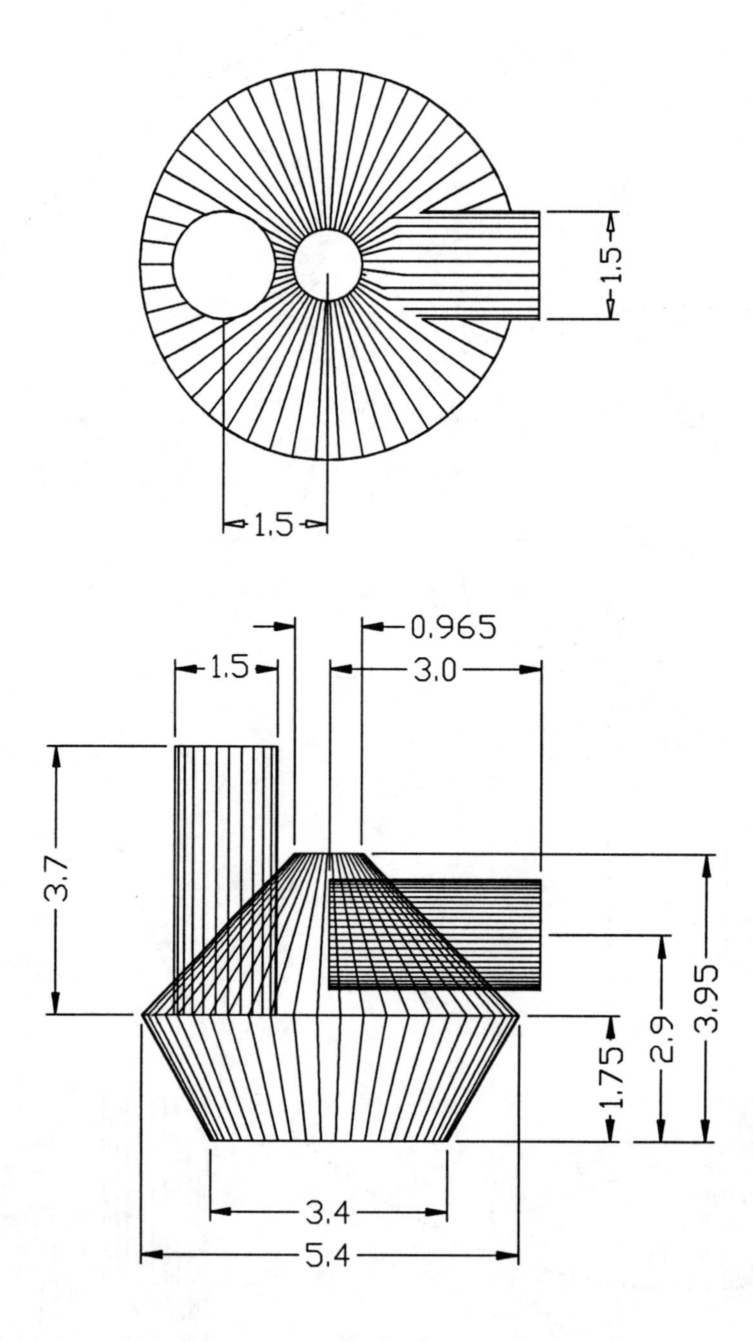

PROJECT-6

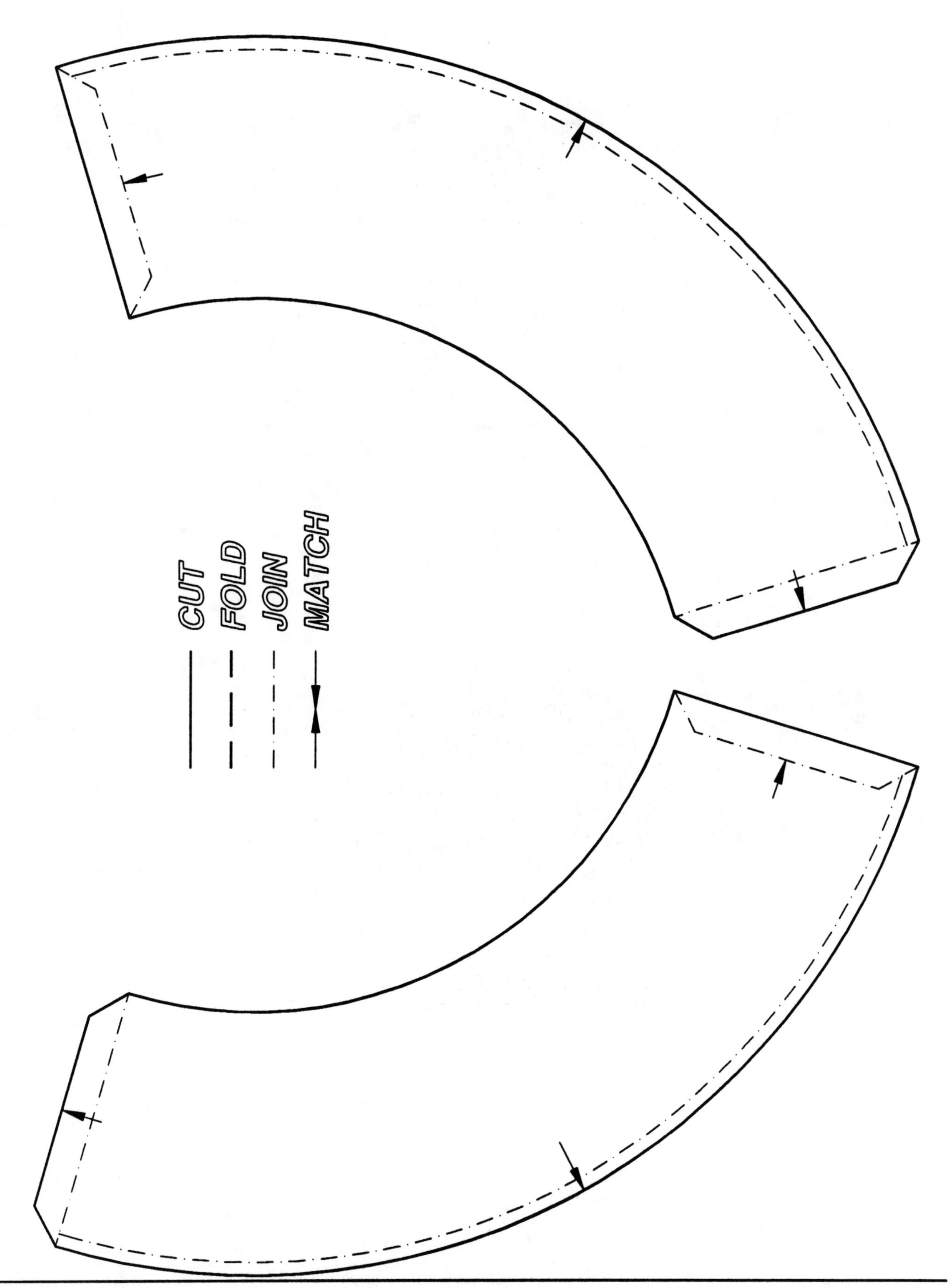

PROJECT-6

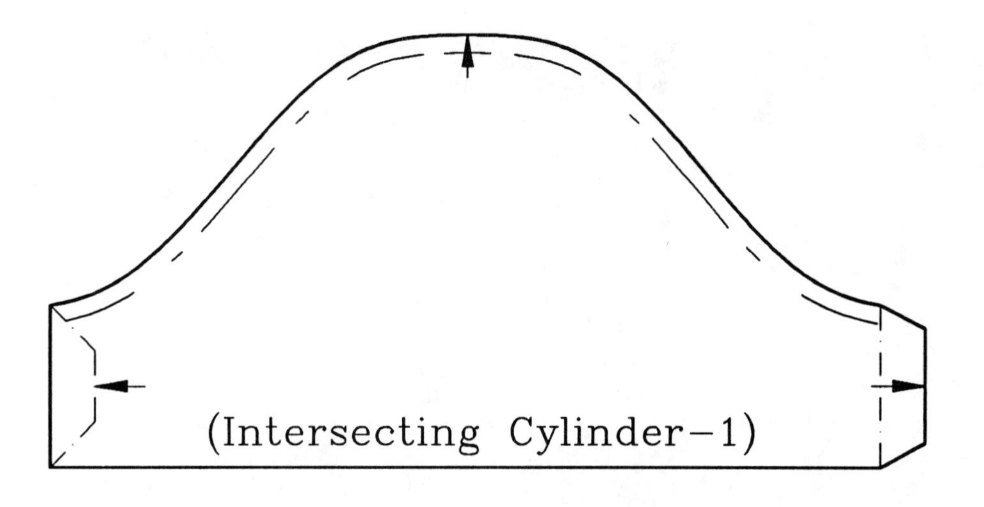
(Intersecting Cylinder–1)

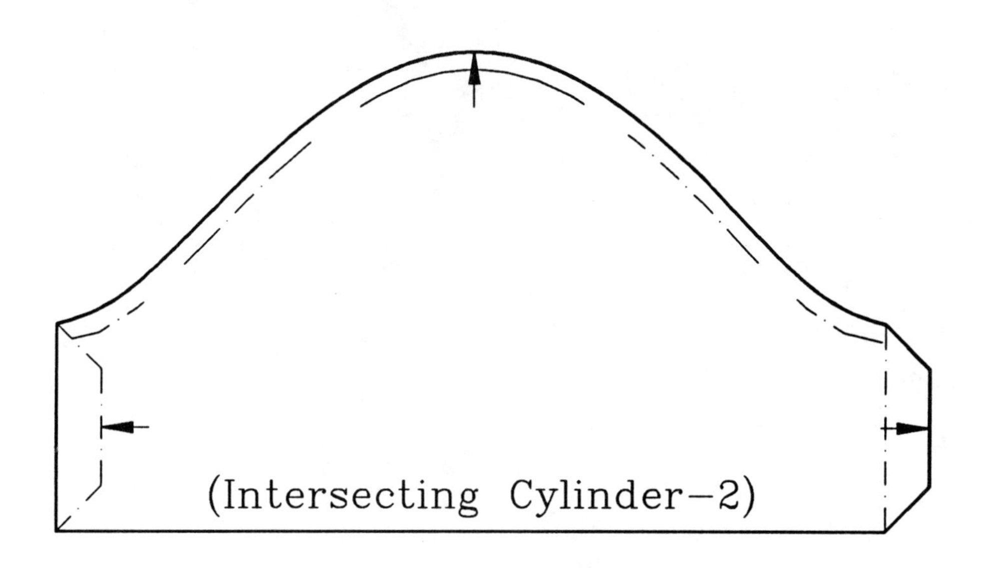
(Intersecting Cylinder–2)

JOIN
MATCH
CUT
FOLD

PROJECT-6

Sheet - 39

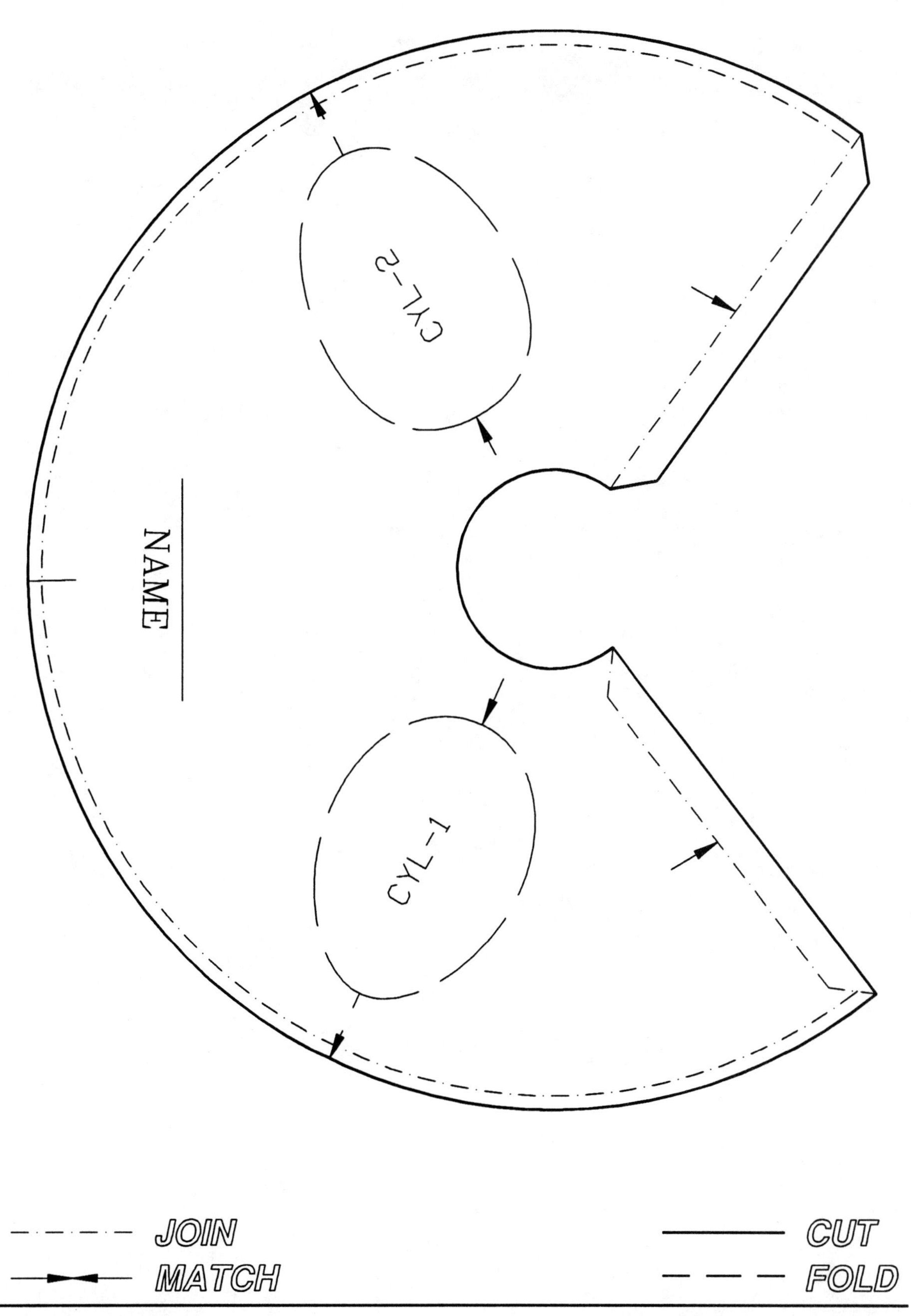

JOIN

MATCH

CUT

FOLD

PROJECT-7

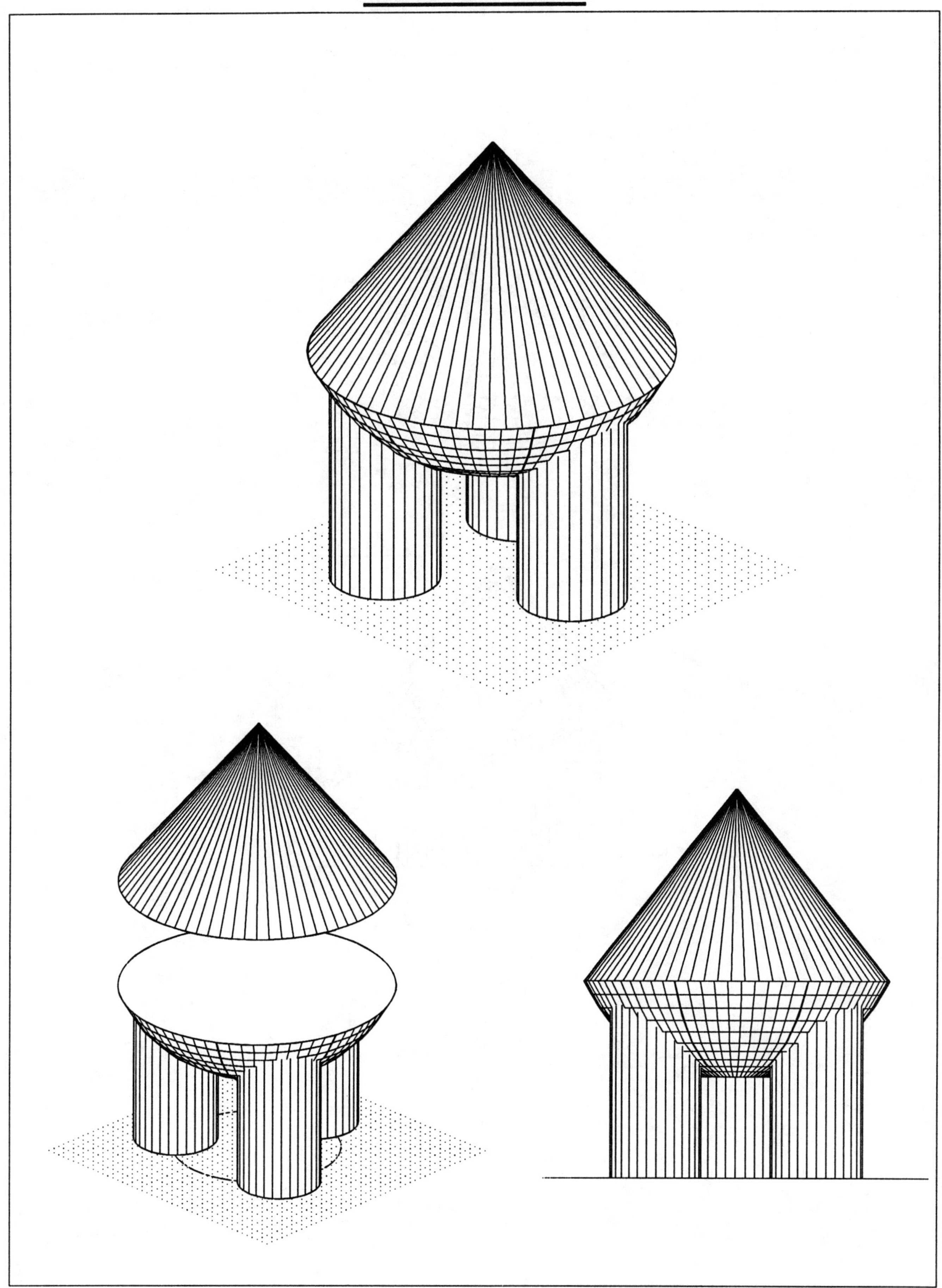

PROJECT-7

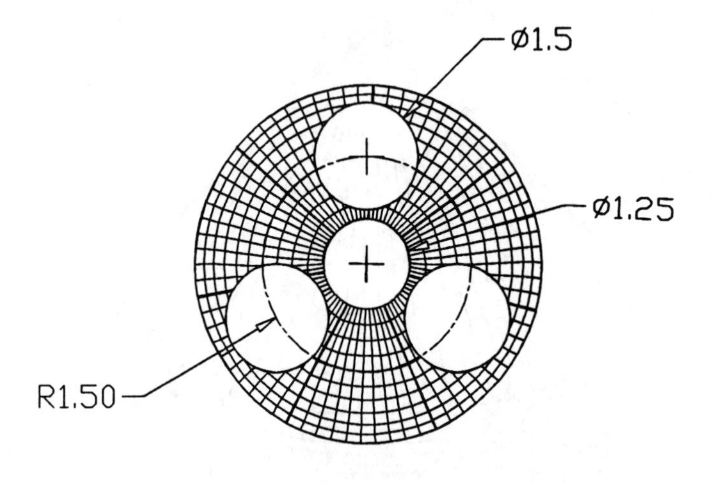

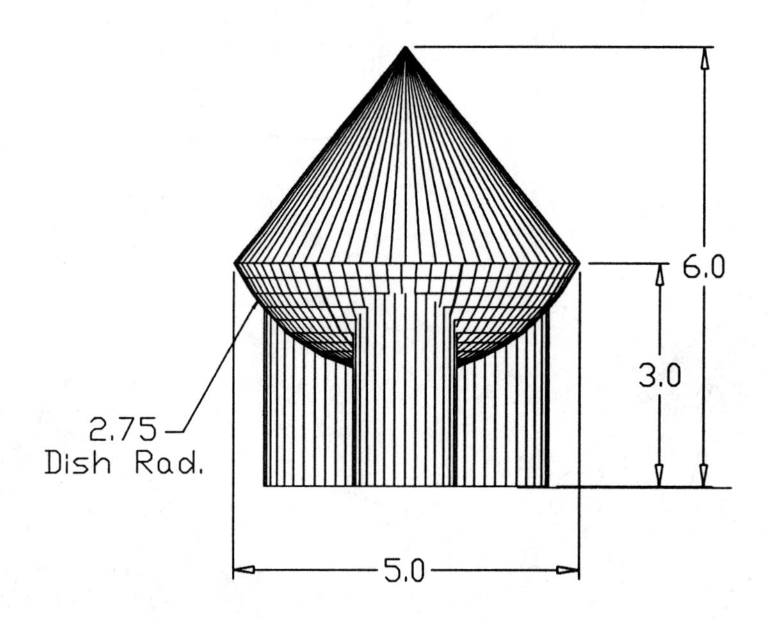

PROJECT-7

Sheet - 41

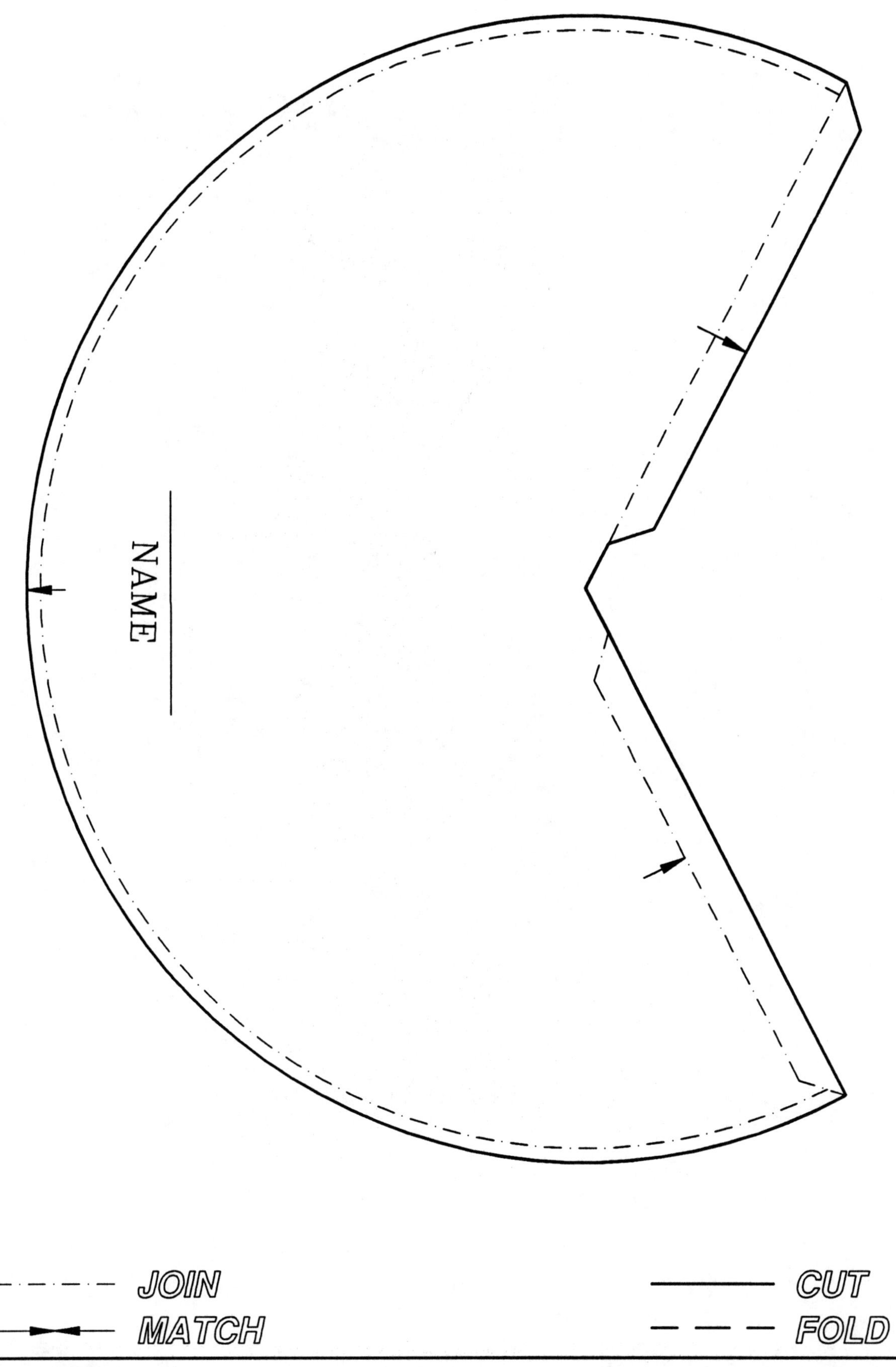

PROJECT-7

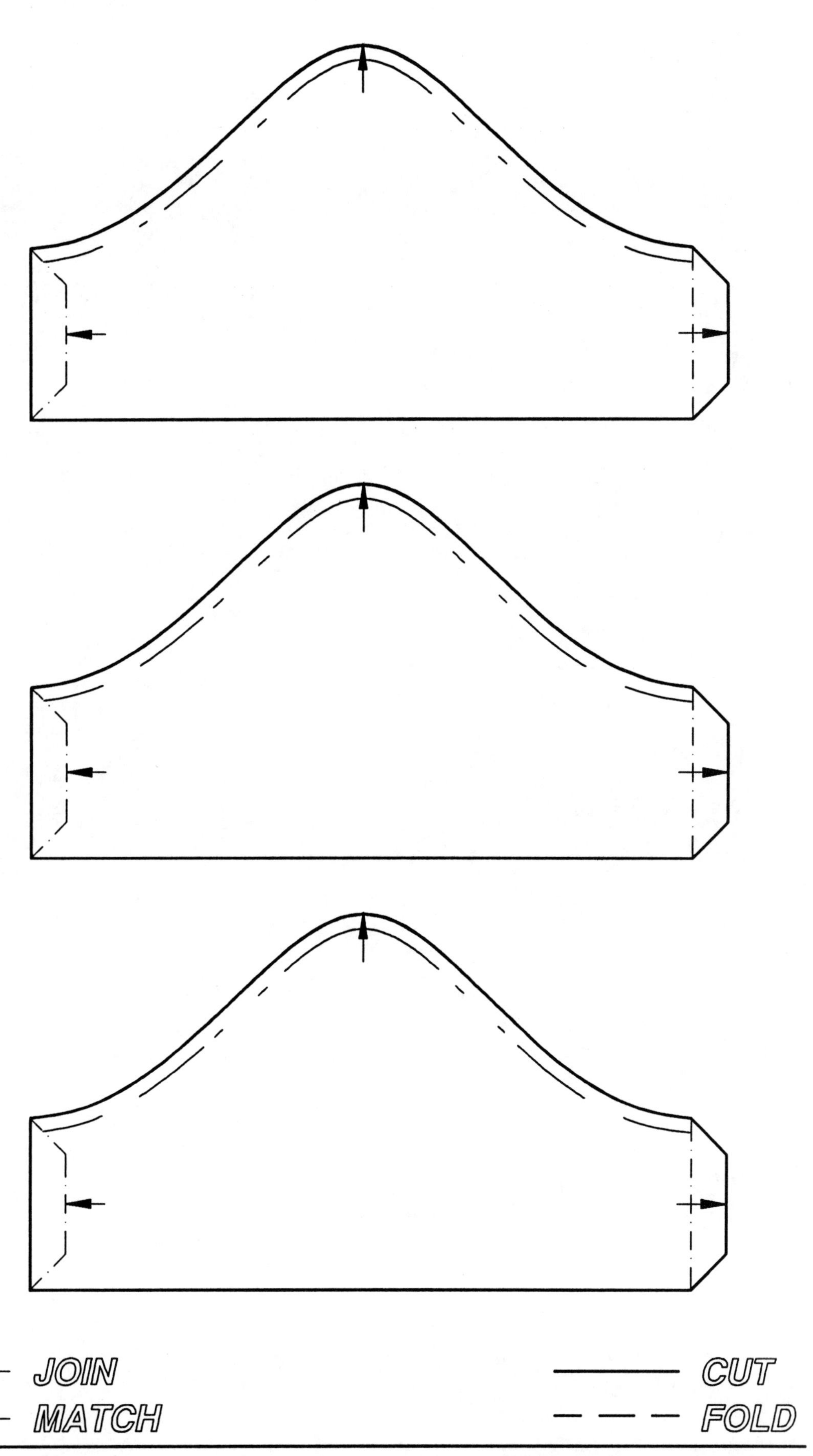

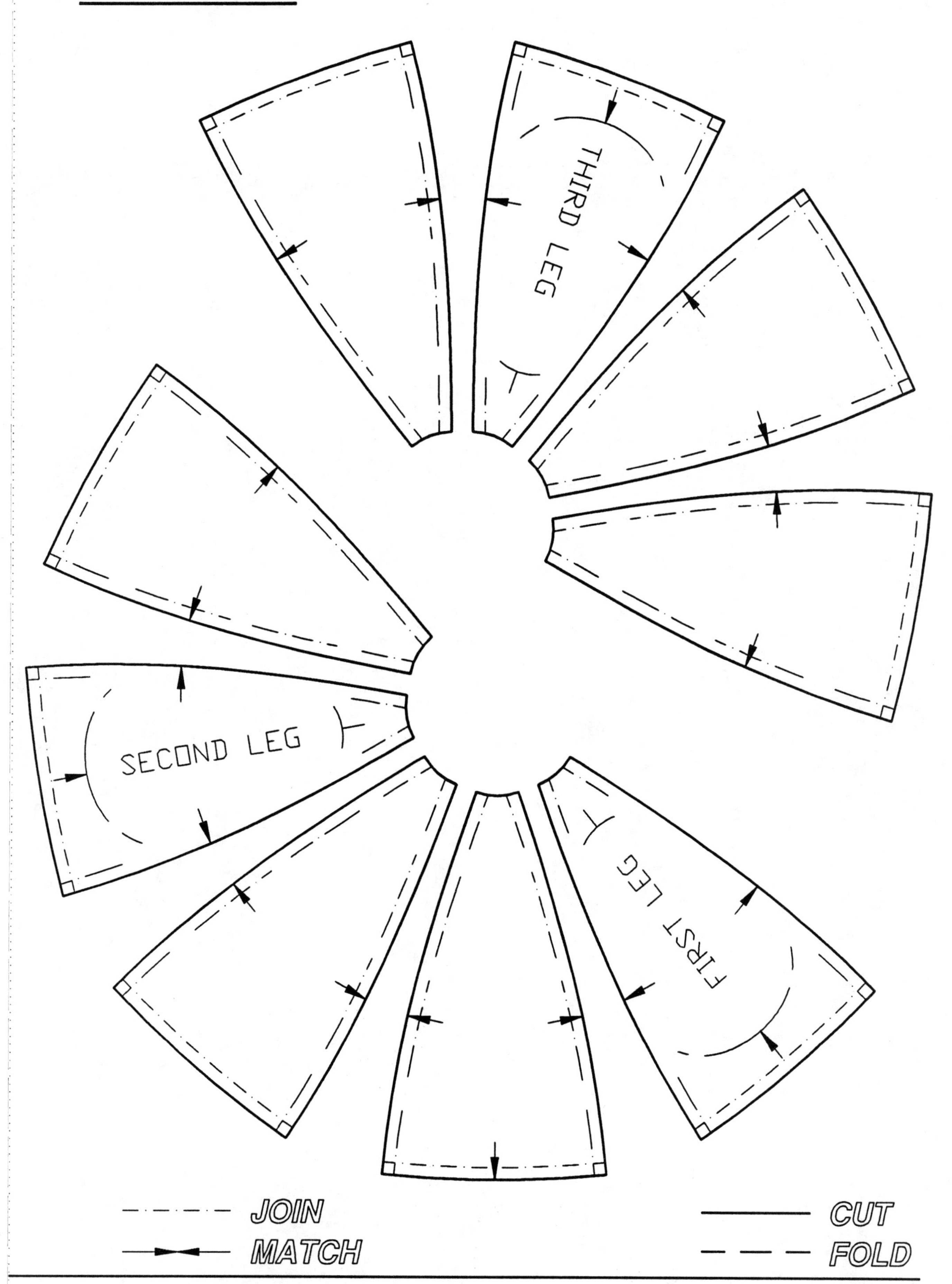
THIRD LEG
SECOND LEG
FIRST LEG
JOIN
MATCH
CUT
FOLD

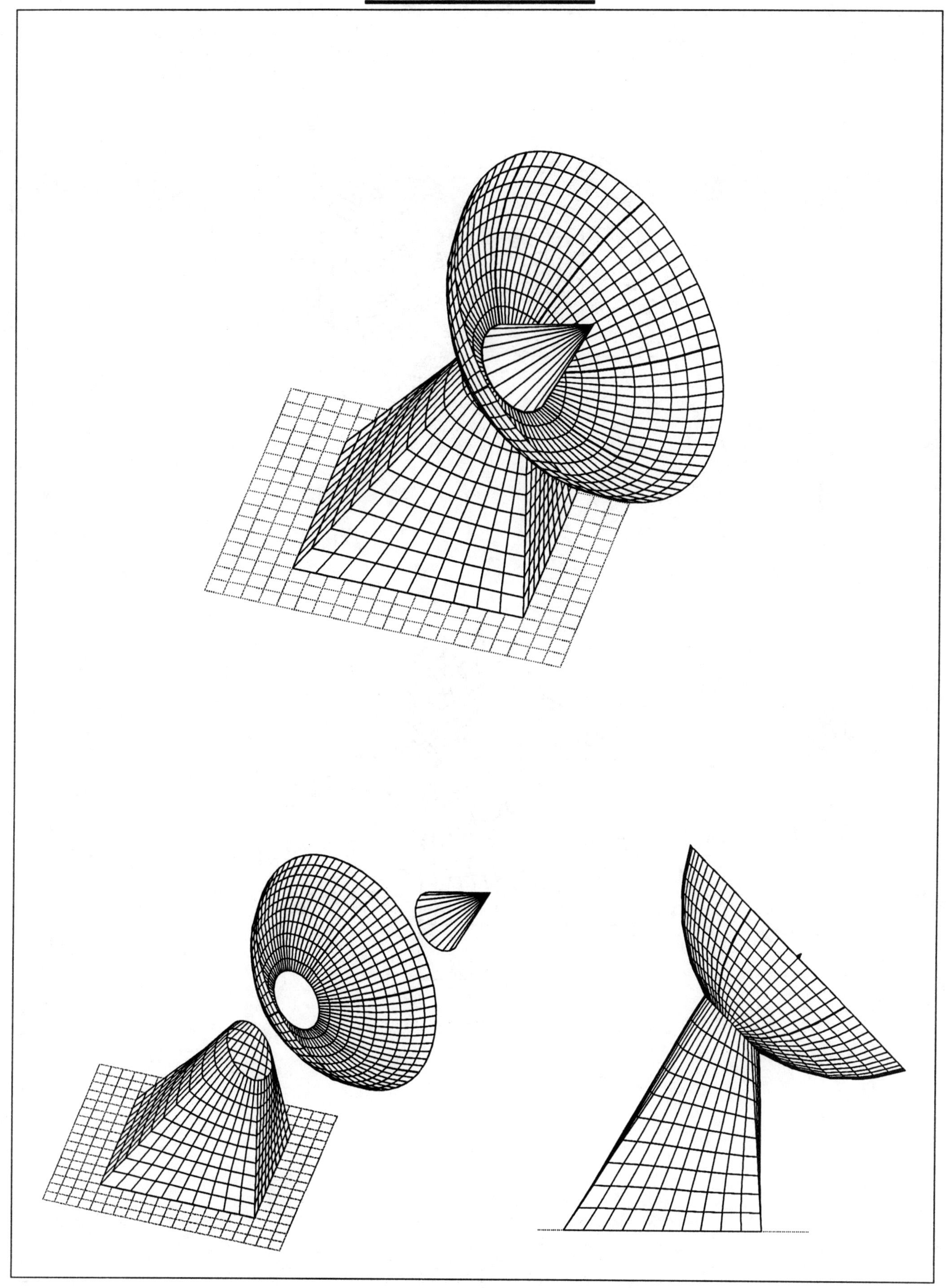

PROJECT-8

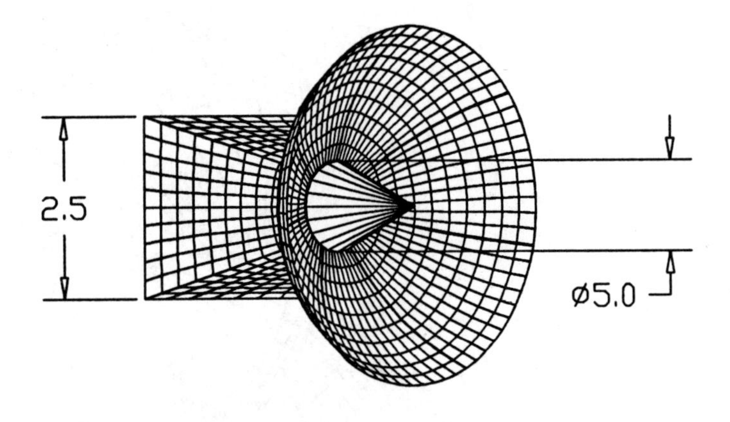

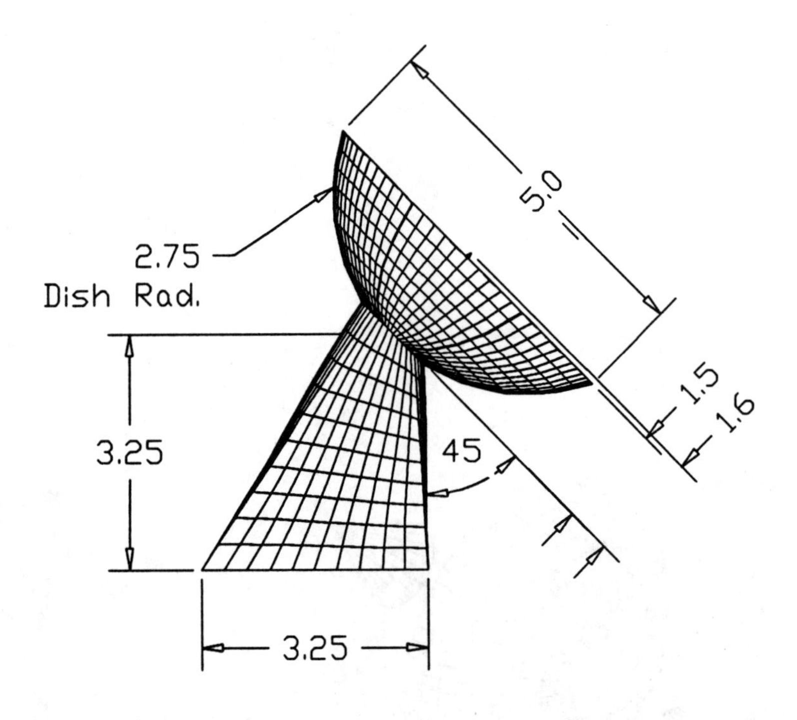

PROJECT-8

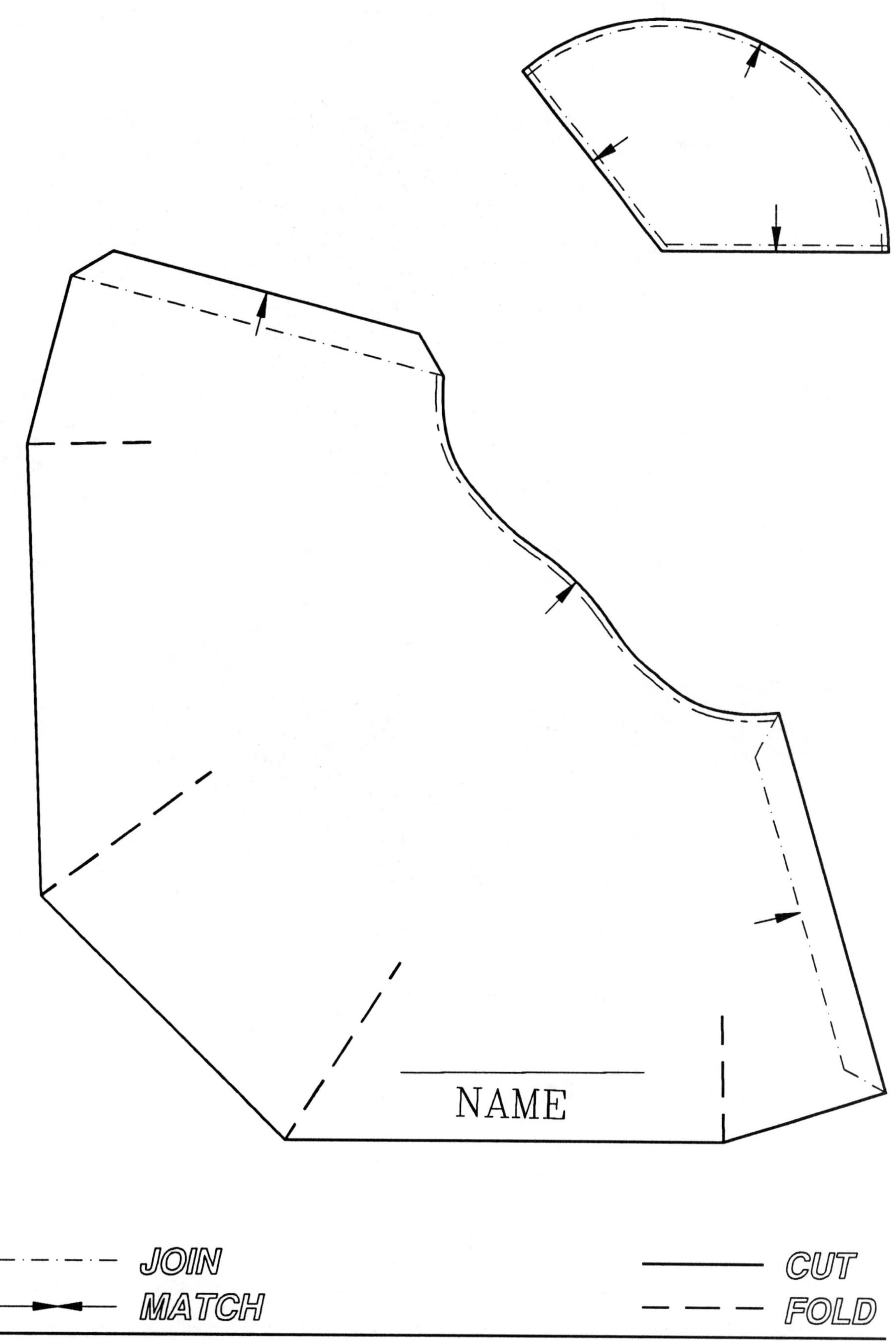

PROJECT-8

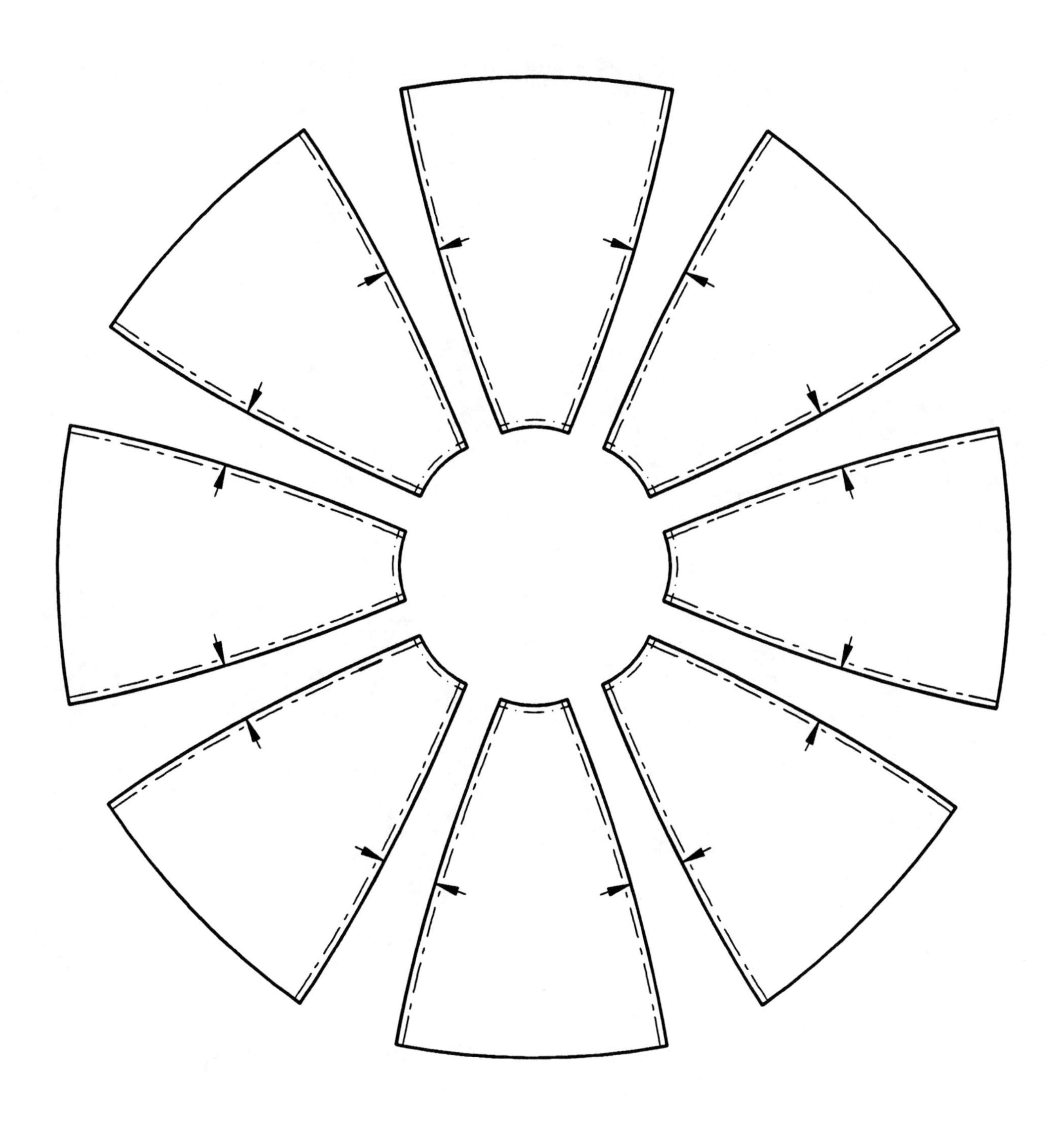

CUT

FOLD